The
Little Scientist
An Activity Lab

The Little Scientist

An Activity Lab

Jean Stangl

TAB | **TAB BOOKS**
Blue Ridge Summit, PA

FIRST EDITION
FIRST PRINTING

© 1993 by **TAB Books.**
TAB Books is a division of McGraw-Hill Inc.

Library of Congress Cataloging-in-Publication Data

Stangl, Jean.
   The little scientist : an activity lab / by Jean Stangl.
     p.   cm.
  Includes index.
  Summary: Describes forty science experiments involving such
subjects as temperature, color, and water.
   ISBN 0-8306-4101-7   ISBN 0-8306-4102-5 (pbk.)
   1. Science—Experiments—Juvenile literature.   [1. Science-
-Experiments. 2. Experiments.]   I. Title.
Q164.S73 1992   92-33743
507.8—dc20   CIP
                                   AC

Acquisitions Editor: Kim Tabor
Supervising Editor: Joanne M. Slike
Book Editor: Annette M. Testa
Indexer: Joann Woy
Director of Production: Katherine G. Brown
Layout: Rhonda Baker
Typesetting: Ollie Harmon
          Susan E. Hansford
          Tina M. Sourbier
Design team: Jaclyn J. Boone, Designer
          Joanne Slike
Cover design and illustration: Denny Bond, East Petersburg, PA   KIDS

To Mary (M.K.H.)

Contents

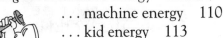

Introduction

Why a lab for young children? The purpose of *The Little Scientist* is to help young children become aware of the world around them and to stimulate their desire to explore, investigate, and experiment—and, thus, make discoveries.

As teachers or parents, you will act as the facilitator and provide the time, opportunity, and encouragement to help young children make these discoveries by introducing them to ways in which they can use both indoor and outdoor environments as learning laboratories.

You and your children will find that science is all around you and the world is a giant laboratory.

As adults, we cannot presume to *teach* science to young children for they *learn* science by making their own discoveries. At this age, they are naturally curious, inquisitive, and eager to find out "how" and "why" things happen. They will become aware of the exciting revelations of science and nature if we provide the materials, equipment, and conducive environment.

By our own enthusiasm and interest, we set the mood for the lab projects. We want to encourage children to take extra time to look closely and to learn to examine objects with a hand lens in order to heighten their observation skills. By making available paper and crayons or felt markers for illustrating or writing down key words, we help them develop the habit of recording their observations.

We can often stir up their curiosity by simply stating, "I wonder what this is?" or "I wonder what would happen if . . . ?" After a few lab experiences, your little scientists will surprise you with their intelligent guesses and questions.

As adults, it is important that we plan our experiments with the children together, share in their joy and excitement of even the smallest discoveries, and help them find the answers to their questions.

The environment

The world, itself, is a playground for learning. Wherever you go, wherever you are—that's the place to begin exploring. The classroom, home, and surrounding outdoor areas all provide excellent opportunities for science discovery. An overturned rock in the play yard, a picnic in the park, or a walk around the block will open up unlimited possibilities.

After a little practice in observing, you and your children will soon find there are numerous surprising discoveries to be made within the small plot of ground right under your feet or on the branch of a tree just above your heads.

We often overlook the science possibilities that are available to us in the classroom and at home. For example, the crayons we draw with, the food we had for a snack, the dead fly in the window sill—all of these are open invitations for lab-type exploration.

By semi-structuring or adding to the environment, you will find you are able to extend the learning opportunities. Children will be quick to notice new items in the lab and will look forward with eager antici-pation to the day's experiments.

With young children, a great deal of learning takes place through the senses. You will want to take advantage of every opportunity to stimulate sensory awareness and to provide additional materials to the lab that will encourage feeling, smelling, listening, tasting, and observing.

Materials and equipment

The tools young children will need for the activities and projects in this book are simple, everyday items. Most are readily available in the classroom and at home. For example, your children can help recycle throwaway items for use in the lab or you can discuss with the children ways in which we can all conserve and help protect our environment.

Furthermore, by making available an assortment of art materials, you will be taking advantage of one of the best ways in which young children express their natural desire to be creative. Their own drawing will reinforce what they learned from the experiment.

The most valuable and necessary tool for your young scientists is a sturdy hand lens. For the classroom, you will want to invest in several hand lenses of various sizes. For home, the little scientists need one that is easy to handle, with a clear viewing lens. Both teachers and parents will find that a small magnifying glass, which fits in pocket or purse, will prove invaluable for closer scrutiny of tiny specimens that often make unexpected appearances.

Activities and experiments

The activities and experiments in *The Little Scientist* are set up so young children can not only become physically involved in the process, they can also do many of the activities on their own or with minimal adult supervision. Those requiring special supervision or potentially hazardous tools are clearly indicated with a symbol next to the corresponding activity.

Research has shown us that young children learn best by *doing* and that they remember and can recall more when they have been actively involved. It is hoped that you, as teachers and parents, will select from this book those activities that are the most suitable for the age and ability of your children. Remember that maximum learning does not take place when children are *watching* an adult do the experiment, so get the children involved.

For each activity you will find background information, the goal, some tips, the materials you need, how to do the lab activity, and some follow-up suggestions of things to think about and discuss.

A visit to the library, local museum, rock shop, seashore, or other field trips can be a valuable extension of the school and home laboratory. By experiencing science as a part of everyday life, children will gain an appreciation of nature that will help keep alive their innate sense of wonder that will soon disappear if not nurtured.

Part 1
Temperature–
Is It Hot or Cold?

Through the activities and experiments in this section, children will learn about temperatures, how our bodies help us determine hot and cold, and about several types of thermometers. They will make a bottle-type thermometer and compare it to commercially made ones. The little scientists will discover why some clothing keeps us warmer than others.

In the first group of activities, little scientists will see how our bodies help us distinguish between temperatures, and that our bodies are sensitive to temperature changes. Children will become aware of tactile experiences with regard to temperature, using the entire body as well as the hands and mouth area; and how our bodies tell us when something is warm or cold.

Children will work together to make a simple bottle-type thermometer, and then test it to see how it works to show rising and falling temperatures.

Children will discover many kinds of thermometers as they investigate and experiment to see how different ones work and what they are used for. Experiences are provided for those used for cooking, in the freezer, and outdoors. They will see how each thermometer registers hot and cold temperatures and will learn about the freezing point.

In the last section, the little scientists will experiment with fabrics and thermometers to learn how warm clothing insulates our bodies against cold. They will have an opportunity to make a temperature prediction and then experiment to see if they were right.

Learning about temperature through the senses

Goal To become aware of temperature, and to be able to distinguish between two temperatures using the senses.

New words Temperature, thermometer

Teacher/Parent tips
- Temperature is how hot or cold something is as measured by a particular scale.
- Air temperature is the degree of heat in the atmosphere. When having children make temperature comparisons, determine where large differences exist.

Lab needs
- Two bowls
- Two clear-plastic drinking glasses
- Water
- Ice
- Two small paper drinking cups for each child
- Fruit juice
- Ice cubes
- Water

 (If a hot-water source is not available, heat water in an electric coffee or teapot. Ice cubes can be kept in a wide-mouth thermos bottle.)

Scientific principles
- The body is sensitive to temperature changes.
- Warm air rises.

Purpose The purpose of the following activities is to show that our bodies respond to temperature. The experiments will also show that our hands, lips, and mouths can help us tell differences in temperature.

3

Children will discover changes in temperature by the way their bodies feel. The heat from the sun coming through the window makes them feel warm, and when they move away from the heat source, they feel cooler.

They will also find that they can tell the difference between the temperature of the water in two bowls by using their hands to feel the water.

By feeling the outside of the glass in another activity, children will be able to distinguish between the temperature of the water inside the two glasses without touching the water.

Their lips and mouths will become an indicator of temperature as they taste warm and cool fruit juice.

By climbing closer to the ceiling, they will be able to feel the warmer air and be able to feel that warm air rises.

Activities
1. Use your body to "feel" temperature by first standing near a sunny window for 15 seconds, and then standing in a cooler area of the room for 15 seconds.

2. Use your hands only to "feel" and determine differences in temperature. Fill one bowl with very warm water and another with cold tap water. Place one hand in each bowl for five seconds to determine which is the warmest.

3. Fill one clear-plastic drinking glass with ice water and one with tepid water. Feel the outside of the glasses to determine which one is the coldest.

4. Determine the coldest juice through a taste test by sampling two small cups of juice—one at room temperature and one that has been cooled with ice.

5. If you have safe climbing apparatus, compare the air temperatures at floor level and at near ceiling level.

Explanation In the first activity, we found that the sun gives off heat, which warms our bodies. When we move to a cooler place, the cooler air around us cools our body. Because dark colors absorb more heat, children wearing darker cloth-

5

ing might feel warmer by the window than those wearing lighter colors.

In the next activity, we were able to use our hands, or our sense of touch, to feel which bowl of water was the warmest. Our feet could have been used in this experiment as well, but our hands are more convenient and useful in determining the water temperature differences in this instance.

With the third activity, we found that the ice water cools off the glass and the cold glass makes our hand cooler. The water in the other glass is about the same temperature as our hand, so we do not feel any change.

Similarly, the fourth activity shows how our lips, tongue, and the inside of our mouths are other parts of our bodies that help us to feel temperature. Along with helping to determine temperature, the taste buds, (little bumps on our tongue), and the inside of our mouths, also help us to taste the flavor of the juice.

Lastly, with the climbing activity, we feel warmer when we are closer to the ceiling because the warmer air moves toward the ceiling. We found it was cooler lying on the floor, because the cooler air stays close to the floor. For the same reason, the person who sleeps in the upper bed of two bunk beds will be warmer than the person sleeping on the lower bed.

Follow-up
- Discuss the results of the experiments and how we are made aware of different temperatures through our senses.
- Explore other ways to find out which glass of water or juice was the coldest.
- Explain that warm air rises and the air at ceiling-level will always be warmer than the air at floor-level.

Learning about making a thermometer

Goal   To discover how to make a tool to measure temperature.

Teacher/Parent tips

+ Explain to the children that in the bottle-thermometer we are going to make, the colored water will rise in the straw as the temperature around it rises.
+ Explain that the red food coloring will make the thermometer easier to read.
+ Be sure to seal the space around the straw thoroughly. If air is allowed in, the thermometer will not work properly.

Lab needs

+ A small nail
+ Hammer
+ A tall bottle or jar with a tight-fitting lid
+ Clear-plastic drinking straw
+ Red food coloring
+ Water
+ Small piece of modeling clay
+ Kitchen timer

Scientific principle

+ Both hot and cold temperatures can be measured by using the appropriate thermometer.

Purpose   The purpose of the following activities is to involve the children in making a simple bottle-thermometer that they can use to observe the change in temperature. They will discover that when they move the thermometer to a very warm place, the red water moves up, and when they move it to a cooler place, the red water goes down. This will help them understand that, as the air temperature around the thermometer warms up, the volume of the colored water expands and rises. The concept will be reinforced by first observing and then drawing a picture to illustrate the experiment. They will also become aware that the movement of the red water in their thermometer is an indicator of temperature changes.

Activities

1. Use the nail to punch a hole in the center of the lid of a tall bottle or jar. The hole should be big enough to insert the clear-plastic drinking straw. Fill the bottle to the top with water and add several drops of food coloring. Replace the lid. Push two-thirds of the straw into the colored water. Press modeling clay around the straw to seal the lid.

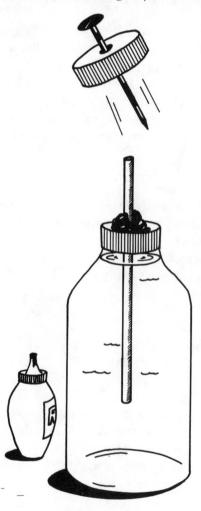

2. Move the thermometer to different areas in the room where the temperature fluctuates (e.g., near a sunny window, by a cool cabinet, close to a heating or cooling source). Set a kitchen timer for five minutes, then check and record the changes. Draw a picture of each location you tested and then make an arrow pointing up or down to indicate the direction the water moved in the straw.

Explanation By coloring the water, we are able to see it more clearly. It was important to seal the hole around the straw where it fits into the lid in order to keep outside air from entering. We also needed to fill our bottle to the top so there would be no air in the bottle.

When we finished making the thermometer, we did not see the red water moving, but when we moved it to a hot, sunny window, we saw the water rise upward. When we moved it in front of the air conditioner (or fan) we saw it move down the straw. Our thermometer told us the air was cooler by the air conditioner. The arrows that were drawn in the pictures show us that in cooler places the water moves down the straw, and in warmer places, it moves up.

Follow-up • Talk about the warmest place and the coldest place found and why this might be so.
  • Ask for suggestions as to where the thermometer could be placed to show more prominent temperature changes, such as in a pan of hot water or in the refrigerator.

Learning about types of thermometers

Goal   To develop an understanding of various types of thermometers and how they work.

Teacher/Parent tips
- A thermometer is an instrument that measures liquids, gases, and solids. Liquid-in-glass thermometers are the most familiar. The red column of liquid is either colored alcohol or mercury. When the temperature goes up, the volume of liquid expands and the liquid rises.
- On the Fahrenheit scale, 32°F is the freezing point of water and 212°F is the boiling point.

Lab needs
- An assortment of thermometers (e.g., oven, freezer, meat, candy, large, easy-to-read indoor and outdoor types, liquid, those with hands, digital types)
- Refrigerator
- Oven
- Two hot dogs
- Glass of hot water
- Table knife
- Clear-plastic drinking glasses
- Food coloring
- Hot water

Scientific principle
- Different types of thermometers are required when measuring hot and cold temperatures that are produced from different sources.

Purpose   The purpose of the following activities is to make children realize there are many kind of thermometers, but they all serve the same purpose—to measure temperature. As they explore the differences and similarities, they will learn which ones are used indoors, outdoors, and in refrigerators and in ovens.

Children will examine a variety of thermometers and discover new ways of using them. They will realize that thermometers come in different shapes and sizes, and that the thermometer's temperature can be read with its hands, red liquid, or digital numbers. The little scientists will develop the understanding that although all thermometers are used to measure temperature, each one has a special use.

By placing a thermometer in a heated oven, they will see that ovens give off heat and the degree of heat is measured on the thermometer. The refrigerator/freezer observations will show that cold temperature will cause the temperature on a thermometer to fall, but not below 32°F (the freezing point). When placed in the freezer, however, the temperature on the thermometer will always stay below 32°F. Children will discover that in both cases the temperature on the thermometer will fall and the thermometer tells them how cold it actually is in each place.

The hot dog test (Activity 4) will show that the inside temperature of a hot dog that has been in hot water will be warmer than a hot dog that has been refrigerated.

They will be able to see the hands on the oven and freezer thermometers move as the temperature registers. In these activities, they will use several kinds of thermometers to determine temperatures and see how each one differs and how they all differ from the bottle thermometer made in class.

By participating in the activities, the children will discover that an oven gives off heat, that a refrigerator and freezer give off cold air, and that both temperatures can be measured with a thermometer.

The activity using the candy thermometer will show how it registers as they test three different temperatures of water.

By testing areas indoors and outdoors, the little scientists will come to the conclusion that all indoor areas will not always register the same temperature, and neither will all areas outdoors. They will know by the temperature that a dark corner is a cooler spot than by a sunny window, and that temperatures vary from shady spot to a sunny one.

As children check their home for various types of thermometers, they will recognize and become more aware of those used by their family, for what they are used, and who usually uses them.

Activities 1. Examine the various thermometers. Explain how the different types work and their uses.

2. Set the oven thermometer in the oven and turn the oven on. Close

the oven door, turn on the oven light and observe the change in temperature on the thermometer.

3. Set the freezer thermometer in the refrigerator with the door open and observe. Wait a few minutes, then record the temperature. Now move the thermometer to the freezer section and compare the two temperatures.

4. Use a meat thermometer to compare the internal temperature of one hot dog that has been refrigerated and one that has been in a glass of hot water for five minutes. Cut the hot dogs into bite size pieces for a taste test.

5. Use the candy thermometer to test different degrees of hot water in three glasses. Add food coloring for easier identification.

6. Purchase easy-to-read thermometers for indoors and outdoors and hang them where they can be easily viewed. Decide on places to test the indoor thermometers (e.g., sunny window, floor, dark corner). Record the indoor and outdoor temperatures at three different times during the day on a chart.

Explanation   Review the uses for the thermometers that were implemented in the activities. Explain that the freezer thermometer measures only cold and the oven, candy, and meat thermometers only measure heat. Remind the little scientists how the numbers on the thermometers used in the kitchen differ from the ones used on an outdoor thermometer, because outdoors we are measuring both hot and cold temperatures. Meat thermometers measure the temperature inside the meat that is cooking and tells us whether it has cooked long enough. Digital thermometers are the easiest ones to read.

Turning the oven on caused it to start heating up. The longer the oven is on, the higher the temperature will go (to a certain point).
As found in activity 4, meat sitting in hot water will be warmer inside than meat taken from the refrigerator. Similarly, we felt a difference in the warmth and coolness of the hot dog when we tasted it.

Discuss the freezing point (32 degrees) that is shown on a thermometer. Explain to the little scientists that in a freezer it is necessary to keep the temperature below the freezing point so food will freeze and not spoil. In the refrigerator, however, explain that the food kept there must not be frozen and thus the temperature should be kept above the freezing point. In activity 5, we could not tell by looking at the three glasses of water which one was the hottest, so we used our candy thermometer to find out. This thermometer is used when making candy to determine how long to cook it.

We can see by testing with a thermometer that different locations, both inside and outside, have different temperatures. For example, we have found that the sun gives off heat and that a themometer measures a sunny area warmer.

Follow-up   • Discuss why it is important to know various temperatures.
• Talk about other places where thermometers might be found.
• Have children think about the times when they were sick and ways that were used to determine if they had a "temperature," such as feeling their forehead and taking their temperature with a thermometer.

13

Learning about fabrics that protect us

Goal   To discover that some materials (e.g., wool, cotton, silk) keep us warmer or cooler than others.

Teacher/Parent tips   • The following activities will help children understand which of the three fabrics would be the warmest to wear on a cold day. However, in the following experiment, the differences in the way the fabric is wrapped around the thermometers and how quickly they are wrapped and unwrapped will affect the outcome.

Lab needs
• A 1-inch square of wool fabric
• A 1-inch square of cotton fabric
• A 1-inch square of silk fabric
• Three rubber bands
• Three easy-to-read liquid thermometers
• A glove or mitten

Purpose   The purpose of the following activities is to help children identify three types of fabric, and to exercise logical thinking skills through their prediction as to which one would keep us the warmest.

     Children will discover the names for the three fabrics, how each fabric feels, and that clothing insulates our body against cold. They will also learn that clothing made from some fabrics will keep us warmer than those made from other fabrics.

     Through experimentation, they will see gloves or mittens protect our hands from the cold.

Activities   1. First examine the fabric squares of wool, cotton, and silk. Then write the words wool, cotton, and silk on a chart. If you had a shirt made from each of these squares of cloth, predict which would keep you the warmest. Write your prediction on a piece of paper.

2. Wrap one square of fabric around the bulb of each thermometer and secure the cloth with a rubber band. Place the thermometers side by side in a cold spot for 10 minutes.

3. Quickly unwrap and record the temperatures. Compare the results with your prediction.

4. Put a glove on one hand. Go outside on a cold day and keep the bare hand uncovered.

Explanation Clothing is made from many kinds of fabric. Some will keep us warmer than others. As we found in activity 2, the thermometer wrapped in wool had the highest temperature. This tells us that if we had a coat made from each of the three fabrics, the wool coat would keep us the warmest.

15

The last activity showed us that if we wear gloves, we can go outside on a cold day and our hands will stay warm. If we lose one glove, the bare hand will get cold while the hand with the glove stays warm. It also showed us that warm clothing protects us from cold.

Follow-Up ◆ Based on the activities, decide which fabrics should be used for clothing to keep us warm on a cold day and which to use to keep us cool on a hot day.

Part 2
Weather–
Rain, Snow, and Ice

Through the activities in this section, young children will learn to be more observant of clouds and the weather. By participating in the activities, they will also learn how to make their own cloud cover chart to reinforce what they observe.

The scientific principles being taught in this section are: All clouds contain small water droplets or tiny ice crystals; water exists in three states; and the freezing point of water is 32°F.

The little scientists will be introduced to the different types of cloud cover and will learn how to identify clouds by name (especially cumulus clouds). In one of the activities, the children even actually make a cloud.

Our young forecasters will soon become more attentive to weather reports from the media as they learn more about the forecaster's job. Through their observations and charts, the little scientists will also see that weather cannot always be predicted.

Studying the water cycle, snow, snowflakes, and crystals found in an ice cube, snowflakes, and in some rocks, will add to the children's knowledge of the weather.

Learning about clouds

Goal To become aware of the change in cloud cover through observations and recordings.

New words Predict, cumulus cloud, stratus cloud, nimbus cloud, cirrus cloud

Teacher/Parent tips
- Weather forecasters observe and study clouds carefully. Certain types of clouds help them predict storms, rain, snow, or sunny weather. As children learn more about clouds, give them an opportunity to make weather predictions.
- Types of clouds: *Cumulus*: piled-up masses of white clouds; *stratus*: clouds that appear as layers; *nimbus*: dark gray rain clouds; and *cirrus*: clouds formed entirely of ice (other clouds are mainly water droplets) that look like small tufts of cotton hanging high in the sky (this type is seldom formed).

Lab needs
- Small bottle lids
- Heavy paper plates cut in half
- Crayons
- Two three inch-long strips of tagboard
- Scissors
- Paper fasteners (brads)

Purpose The purpose of the following activities is to help young children learn about different types of clouds, to observe daily cloud cover, and to make a prediction about the next day's cloud cover. They will have an opportunity to see a weather report on television and to make a chart to record today's cloud cover and to use it to predict tomorrow's cloud cover.

Children will discover that cloud cover can be recorded over a few days by making a chart and over a month's period by using picture recordings on the calendar.

Activities

1. Watch a television weather report (or videotape a program).
2. Observe and describe clouds over a period of three days. Draw attention to cumulus clouds.
3. Use the bottle lid to trace four small circles on the back side of a paper plate. Color in the circles to represent no clouds, slightly cloudy, half cloudy, and full-cloud coverage. Cut the ends of the two tagboard strips to form arrows. Write "yesterday" on one and "today" on the other. Attach the straight ends of the arrows to the bottom center of the plate with paper fasteners. Check the cloud cover each day and reset the hands.
4. After a few days, add a "tomorrow" arrow to use in predicting the next day's cloud cover.
5. Trace, color and cut out several sets of clouds. Each day, check the cloud coverage, find the corresponding circle, and attach it to a calendar.

Explanation Weather reports give us an idea of what the coming weather will be like. They tell us about storms, rain, snow, and cloud cover and predict the coming temperatures.

As you observe clouds, point out the differences and similarities and give the name of the types of clouds seen (see Teacher/Parent tips above). For example, the white fluffy clouds that look like piles of white cotton are the cumulus clouds.

The weather charts drawn in activity 3 have helped us in keeping a

19

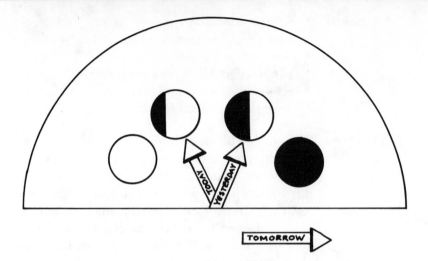

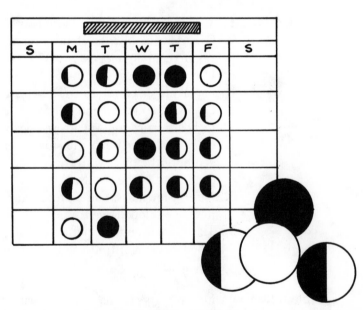

record of the cloud cover for yesterday and today and assisted us in predicting what it might be like tomorrow. Sometimes we were wrong and we learned that weather cannot always be predicted. We can look at our calendar and see how many cloudy and sunny days we had during the month.

Follow-up • Talk about the job of weather forecasters, some of the tools of their trade, and what was discovered from watching the program.

- Discuss each day's cloud cover; how today's cloud cover is different from yesterday; and the reasons for the predictions of tomorrow's cloud cover. Try to think of other ways to predict the weather.
- At the end of the month, remove the cloud circles from the calendar, sort the circles by color, count, and then discuss what was discovered about the cloud cover during the month.

Learning about making a cloud

Goal To discover how clouds are formed.

Teacher/Parent tips • Clouds form when warm air, rising from the ground, cools. The water vapor in the air then condenses and water droplets form. These droplets, when close together, form a cloud.

Lab needs • Two cups of very hot water
• A 2-cup Pyrex measuring cup
• A five-inch-long square of old nylon stocking
• Rubber band
• 6–8 ice cubes

Scientific principle • All clouds contain either small water droplets or tiny ice crystals.

Purpose The purpose of the following activities is to give the students a better understanding of how clouds form by making a cloud in the lab. As they observe cumulus clouds, they will have an opportunity to express their observation by describing the way the clouds look to them. Expressive and descriptive language will come into play.

Children will discover, through the experiment, that clouds form when air, rising from the ground, cools. They will see drops of water forming inside the cup as they watch water vapor rise above the cup to form a tiny cloud. The tactile sense will again be experienced as they compare the feel of the cup before and after it was filled with water, and as they pile the ice cubes on top of the stretched stocking.

Activities **1.** Explain that we need clouds for it to snow and rain, and how clouds are formed.

2. Have an adult carefully pour the hot water into a measuring cup. Carefully feel the outside of the measuring cup both before and after the water is added. When the cup becomes hot, pour out all but one

inch of the water. Stretch the nylon square of fabric over the mouth of the cup and secure it with the rubber band. Pile the ice cubes on top of the stretched stocking. Observe. Water vapor will rise above the cup forming tiny clouds. Drops of water will form on the inside of the cup. When the drops become heavy, they will run down the cup into the water.

3. When the sky is filled with cumulus clouds, lie on your back and study the clouds. Take turns describing one of the clouds.

Explanation   From our experiment of making a cloud, we see how the clouds in the sky are formed. We poured hot water into the cup and left some in the bottom. This warmed the cold air around the cold ice cubes and caused the water vapor to rise above the cup. It formed a cloud that holds the water vapor. The drops of water inside the cup are like the water in the clouds. When the drops became heavy, they ran down the cup. When the clouds become heavy with water vapor, they fall to the earth as rain.

When we looked at the cumulus clouds in the sky, we saw many sizes and shapes. Sometimes we see colors in the clouds. There are people all over the world whose special job is to study the clouds.

Follow-up   • Compare the way the outside of the empty cup felt to the way the cup felt after it was filled with water. Discuss and compare observations and explanations of the experiment as given by each child.

Learning about rain—
the water cycle

Goal   To discover how water is recycled.

New words   Cycle, evaporation, water vapor, rain

Teacher/Parent tips
- Water exists in three states: solid (ice), gas (vapor), and liquid (liquid water).
- Water cycle: The earth's oceans are the main source of rain. The heat of the sun evaporates the water into the air where it remains as invisible vapor until it condenses into droplets of liquid water forming clouds. When the droplets of water become heavy enough, they fall to earth as rain. This process is repeated over and over again. It is the warmth from the sun that lifts water (evaporation).

Lab needs
- Glass jar with lid
- Ice cubes
- Metal cake pan
- Paper
- Felt pens
- Cotton balls
- Glue

Scientific principle
- Water exists in three states.

Purpose   The purpose of the following activities is to help children understand the water cycle and about the three states in which water exists. This will be presented as an adult draws a picture of the cycle. Students should conduct the experiment, then draw their own picture of the water cycle.

    Children will discover from the drawings that water collects in lakes and oceans, heat causes the water to rise in the form of water vapor, clouds form and become heavy, and the water falls back into the lakes

25

and oceans. By participating in the experiment they will have a visual concept of the water cycle.

Activities **1.** On a chart, explain and illustrate the water cycle (i.e., an ocean, vapor rising, clouds forming, rain falling back into the ocean).

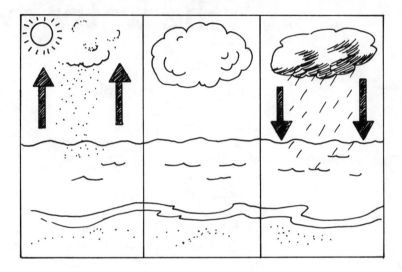

2. Fill a glass jar with ice cubes and put on the lid. Set the jar in a metal cake pan. Observe.
3. When the ice melts, set the pan of water in the sun.
4. Draw a picture to show the water cycle. Glue cotton balls in place for clouds.

Explanation   Review each step of the water cycle from your drawing as children follow their drawings. Compare the ice cubes cooling off the air around the jar, becoming larger (heavier), and then falling into the pan with the air cooling around the cloud, becoming heavy, and then falling as raindrops.

By then placing the pan in the sun, we discover that the sun evaporates the water or draws it up into the air as water vapor just as it does from lakes and oceans. And where is the water (vapor) now? The water vapor is in the clouds.

Follow-up   • Discuss the part of the experiment that involved water vapor and condensation, and where the water in the pan came from.
   • Name some other water sources where evaporation takes place.
   • Review the water cycle.

27

Learning about measuring rainfall

Goal   To discover how to measure the amount of rainfall in a given area.

New word   Rain gauge

Teacher/Parent tips
- A rain gauge is an instrument used to measure the amount of rain that falls in a certain place during a given period of time.
- The rain gauge made in the activities below is similar to ones used by the National Weather Service, but will not be an accurate measurement tool.
- Measurements should be taken during a good rain and can be for any time period (e.g., 10 minutes, an hour, for the duration of the rain).

Lab needs
- Paint stirring stick that fits into the straight-sided jar
- Colored, waterproof marking pens
- Easy-to-read, 12-inch ruler
- Three deep pans (the same size)
- A funnel (the mouth area should be about 10 times that of the jar but needs to set on top of the jar)
- A straight-sided jar 10–12 inches tall

Scientific principle
- Rain can be measured with an instrument called a *rain gauge*.

Purpose   The purpose of the following activities is to have children take measurements of rainfall under different conditions and to make a rain gauge to show how this device measures rainfall more accurately.

Children will discover that the amount of rainfall in each pan, when conditions are kept constant, will measure about the same, while the amount of rainfall in the pans set in three different areas will all differ.

Activities   1. Prepare the paint stirring stick as shown. Each inch should be a different color and should be divided into tenths. Ten inches measured in a straight-sided jar will equal one inch of actual rainfall.

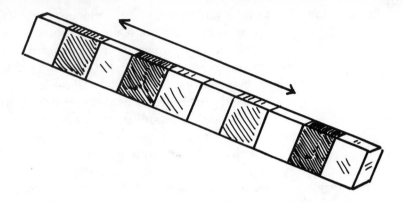

Explain how the rain gauge works (by measuring the water with the measuring stick that accumulates in each deep pan) and show the children the pans. Each time you do the experiment, set the pans out at the same time and bring them inside to measure so as to keep conditions as constant as possible.

2. Set the three pans outside in an open area. Observe. Bring them inside, measure, and record.

Empty the pans and set one under a tree, one under the eves of a building, and one in an open area. Observe. Bring them inside, measure, and record.

3. Insert the funnel into the straight-sided jar and set the gauge in an open area (the jar can be anchored by setting it in a pan filled with rocks or sand). Observe; measure with the measuring stick, then with the ruler; and record.

Explanation We found a way to measure the amount of rainfall by catching rain in three pans. But this did not really tell us how many inches of rain fell.

We also made a rain gauge and discovered another way of measuring the amount of rainfall. Rain gauges gives us the best measurement of rainfall because they are a special tool made for this job. Our rain gauge is a lot like the ones used by people at weather stations.

The formula used to measure the rainfall can be explained by using a piece of paper to represent 1 inch. Cut it into 10 equal pieces to show that 1 inch equals one-tenth of an inch of rainfall.

Follow-up
- Discuss observations and compare recordings.
- Decide which of the three places was the best place to set the pans to measure rainfall, and why.
- Ask for an explanation of how the rain gauge works.

Learning about ice

Goal To discover the attributes of ice.

New word Ice

Teacher/Parent tips
- Low (cold) temperature causes ice to form. Ice is frozen water. The freezing point of water is 32°F. (0°C [Celsius]). Snow, sleet, frost, and hail are ice.
- As water freezes into ice, it expands and increases in volume.
- Heat changes solid ice into a liquid. Cold changes liquid water into solid ice.

Lab needs
- Plastic dishpan
- Ice cubes
- Two clear-plastic drinking glasses
- Two ice cubes (same size)
- Two kitchen timers
- One refrigerator
- One freezer

Scientific principle
- Water freezes at 32°F.
- When water freezes, it expands.

Purpose The purpose of the following activities is to help children learn more about ice; for example, ice is frozen water, cold changes water into a solid (ice), the freezing point of water, water expands when frozen, and heat will change ice back into a liquid (water).

Children will discover that ice melts faster in the sun than in the shade. They will also have an opportunity to make predictions and to experiment to find out if they are right. By experimenting they will be able to prove that water will not freeze in the refrigerator and the amount of water remains the same even after freezing.

Activities **1.** Fill a plastic dishpan with ice and allow time for examination. Give each child an ice cube and have them tell what they know about ice. Make a chart of words that the children use to describe ice.

2. If in a classroom situation, divide the children into two groups: sun and shade. Give each group a clear-plastic drinking glass containing one ice cube. Let each group estimate the time they think it will take for their ice cube to melt. Set the number of minutes on each timer, place the glasses in the given spots and observe.

3. Review temperature and explain how the thermometers work. When the ice cubes in activity 2 (above) have melted, add water to make the water level in both glasses equal. Mark the water levels with a red marking pen. Place one glass in the refrigerator and the other glass in the freezer. Wait 10 minutes, check, and again mark the water and ice levels using a blue marking pen.

4. Predict what we will discover if we set both glasses from activity 2 in a warm place. When the ice melts, again check the markings on the glasses.

Explanation Read together the descriptive words about ice from the chart in activity 1. We were able to check out our predictions by watching the ice cubes melt and discovered that the ice cube in the sun melted first. This shows us that the sun gives off heat and that a sunny place is warmer than a shady place.

Activity 3 showed that water will not freeze in the refrigerator but it will in the freezer because the temperature is colder. When the water froze, we saw that the ice was above our water marking. This proves that water expands when it freezes. But this did not mean we had more water, for when we set the ice in the sun, it melted and showed us we had the same amount of water with which we started.

Follow-up • Talk about how ice looks, feels, and tastes, and some of the uses for ice.
• Discuss the predictions and results of activity 4, and what we discovered from this experiment.
• Discuss what happened when we set the two glasses in a warm place again.

Learning about icicles

Goal To discover how icicles form.

New word Icicle

Teacher/Parent tips
- Icicles are formed by the freezing of dripping water. If it does not freeze where you live, icicles can be made on a smaller scale in a freezer (set the coffee cans on blocks that are high enough from the underlying pans so that good-sized icicles can be formed).
- Bring in icicles for observation or show pictures of hanging icicles.
- The hole that was punched in the coffee can allows water to drip through, a drop at a time. After some water has frozen to the can, more drops flow out and freeze also. The icicle grows from top to bottom with each new drop adding to the ones before.

Lab needs
- Hammer and nail
- A 3-, 2- and 1-pound coffee can
- Heavy string
- Thumbtack
- Red, blue, and green food coloring
- Measuring stick

Scientific principle
- If dripping water freezes, an icicle will form.

Purpose The purpose of the following activities is to introduce children to icicles, to explain what they are made of, and to show how they form. Observing, measuring, and recording on a picture graph will help little scientists learn more about icicles.

Children will discover that when the outside temperature is at or below the freezing point, they can make an icicle. Watching dripping water freeze and form a colored icicle will be an exciting experience. Varying the size of the bottom hole in the coffee cans will show our little scientists that they can create icicles of different sizes. We also learned how to make the holes in the cans. When we hung the cans, we found that it was important for the three pieces of string to be of equal length.

Activities

1. Use the nail to punch three holes, equally spaced, along the top edge of each coffee can. Tie a string through each hole and then tie the other ends of the string together to form a loop with which to hang them. In the bottom center of each can, make a tiny hole using the thumbtack. Vary the size of the holes in each can. On a cold night, take the cans outdoors and fill with water. Tint each can of water with a different color of food coloring. Hang the cans in a safe place.

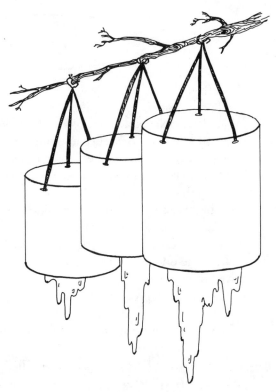

2. The next morning you will find a surprise—shimmering pink, blue, and green icicles! Observe, measure, and record the length of each icicle.
3. Make a picture graph showing the length by color.
4. Put one of the icicles on the table, one in the freezer, and the other outside. Predict what will happen to each one after 30 minutes. Watch the clock and after 30 minutes, check the icicles.

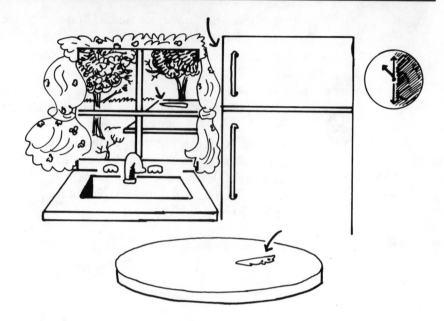

Explanation When the icicles were checked in the morning, we found three icicles of different lengths. This is because the holes in the bottom of the can were each a different size. The larger hole let larger drops of water drip through, making a longer icicle. We also found the icicles were the same colors as the colored water.

The picture graph that was drawn in activity 3 shows the results of our experiment.

In our test and predictions, we found that the icicle on the table melted, the one in the freezer stayed the same, and the one left hanging stayed the same because the temperature outside was still below freezing. It was easier to predict what would happen to the icicles we brought inside, but since we did not know if the outside temperature would stay below freezing, it was harder to predict what would happen to that one.

Follow-up • Discuss what happened when we filled the cans with water, and why there was a difference in the length of the icicles.
 • Share ideas on how the icicles formed.

Learning
about snow

Goal To discover the water content of snow.

New word Snow

Teacher/Parent tips
- Snow is formed when water vapor changes directly into a solid. The formula for measuring the water content of snow is: 12 inches of unpacked snow equals about 1 inch of water.

Lab needs
- A large coffee can
- Measuring stick
- Red, yellow, blue, and green waterproof marking pens

Scientific principle
- The same amount of packed and unpacked snow when melted, will not yield the same amount of water.

Purpose The purpose of the following activities is to help little scientists see that snow is made up of water and to learn one way of measuring the water content of a container of snow.

Children will discover that a can of unpacked snow, when melted, will produce less water than the same size can of packed snow. They will be able to see when they measure how little water there is after the snow melts.

Activities
1. Fill the can to the top with *unpacked snow*.
2. Predict the water level that will result from the melted snow by marking the outside of the can with a red line. Observe. When all the snow is melted, mark the water level on the can with a yellow line.
3. Explain the water content formula, found under "Teacher/Parent tips." Measure the height of the can and the water level with a measuring stick. Record the measurements on a chart.
4. Repeat the experiment filling the can with *packed snow*, this time using the blue and green marking pens.

Explanation We were able to measure both snow and water with the measuring

stick. The activities showed us that the can of packed snow contained more water than the can of unpacked snow. This is because there were more airspaces between the snow flakes in the unpacked snow, thus, there was less water.

By measuring, we found out that if we let 12 inches of packed snow melt in our can, we will only have about one inch of water. From this, we can see that if we wanted to melt snow for drinking water, we would need 12 cans of packed snow to make one can of water.

Follow-up ♦ Discuss and compare predictions. Compare predictions to the results. Compare the two experiments.

Learning about snow and ice crystals

Goal To discover that crystal formations can be found in both snow and ice.

New word Crystal

Teacher/Parent tips
- Both snow and ice are made up of crystals. A snowflake has six sides (a hexagon), is transparent, and has its own design (probably no two are alike). The crystals in an ice cube also have six sides.

Lab needs
- Black construction paper
- Magnifying glass
- Onion skin or other white, lightweight paper
- Scissors
- Ice cubes or icicles
- Broken rocks containing crystals

Scientific principle
- Both snow and ice are made up of six-sided crystals.

Purpose The purpose of the following activities is to help youngsters learn about the crystal formations in snow and ice and to aid them in recognizing crystals in rocks.

Children will discover that snowflakes have six sides, are transparent, and each one has a different design. By catching snowflakes in their open mouths, they will discover that snowflakes are cold, have no taste, and are made up of water.

After cutting the two snowflakes, they will be able to make a comparison that will give them a better understanding of how snowflakes look.

Activities
1. During a snowfall, open your mouth and catch a snowflake.
2. Catch snowflakes on a sheet of black construction paper. Quickly examine them with a magnifying glass.
3. Have each child free-cut a snowflake from a piece of white paper. Compare snowflakes and count the sides.

4. Give instructions for folding and cutting a snowflake as shown in the illustration. Follow the directions and cut a new snowflake.
5. Break an ice cube or icicle. Examine the broken ends with a magnifying glass.

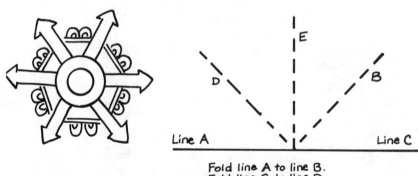

Line A   Line C

Fold line A to line B.
Fold line C to line D.
Fold on line E.
Cut all three sides as desired.
Unfold.

6. Look at the cut side of rocks with crystal formations inside. Look at the varying shapes and sizes of crystals.

Explanation When we tasted snowflakes we found that they had no taste but felt cold to our mouths and tongues. What happened to the snowflakes when they hit our tongues? They melted. This is because the tongue is warm and snow melts in a warm place. Feel your tongue and see if it feels warm.

The snowflakes melted quickly on the black paper, too, but if we were quick enough to see them under the magnifying glass, we saw they had six sides and we were able to see through them.

As we cut our paper snowflakes, we saw that we had to fold and cut correctly to make a snowflake with six sides. Our assortment of snowflakes is like real ones—no two are alike.

When we examined the ice cube, we found crystals. Each ice crystal has six sides.

In looking at our special rocks, we also found tiny crystals, but since rocks do not melt, we can still see the crystals.

Follow-up
- Share what you discovered about snowflakes and how snowflakes taste.
- Take a vote as to whether or not you were able to see and count the sides of a snowflake by using the magnifying glass.
- Compare the snowflake made in activity 3 with those made in activity 4.
- Compare the snowflakes made by class members and see if you can find two alike.
- Discuss the crystals found in the rock, such as how they looked, what colors they were, and how they compare with snow and ice crystals.

Part 3
Colors–The World Is a Rainbow

In the section on mixing colors, children will have an opportunity to experiment by mixing colors to create new colors and to see how they can lighten and darken a color. They will use the new colors for creating a design.

Two ways to change the color of something will be introduced through a dyeing and fading experiment. Paper and cloth will be used to show another example of how color is absorbed.

Projects using crayons will help young artists understand the science principles that heat from the sun can change a solid into a liquid and the liquid can be changed back into a solid by removing it from the heat source.

Upon completing the activities in this part, youngsters will have discovered how to make a rainbow and what the order of the colors are in a rainbow.

Learning about mixing colors

Goal To discover new color combinations.

Teacher/Parent tips
- Children should wear painting smocks and table surfaces should be protected.
- Remove the lids from the egg cartons.
- Cut egg cartons in half, if desired.

Lab needs
- Styrofoam egg cartons
- Powdered art paint
- Water
- White paper towels
- One-half teaspoon measuring spoons

Scientific principle
- When two or more colors are mixed together, the basic color changes.

Purpose The purpose of the following activities is to show how mixing two or more colors will change the original colors, and how new colors can be made when colors are mixed.

Children will discover they can make a variety of new colors from the basic colors and, how they can lighten and darken colors of paint.

Activities
1. Mix equal amounts of two or more colors of powdered paint into each egg cup of the styrofoam egg cartons. Add enough water to make a watery paint.
2. Fold the paper towel three or four times. Spoon the paint onto the folded towel. Unfold and allow to dry.
3. Add three measuring spoons of white paint to one of the paint cups and mix well. Add three measuring spoons of black paint to one of the paint cups and mix well.

Explanation In mixing two colors together, we discovered a new color. You were able to make many new colors using just a few colors. When we added water, some colors looked darker and some colors looked lighter than the dry paint.

45

When we spooned the paint onto the folded paper towel, we could watch the colors soak through the towel and run together to make new colors. As the paper towel was unfolded, more than one design was found, but all the designs were the same. This is because the towel was folded and the paint soaked through all layers of the towel in the same way. The towel absorbed the colors.

Adding white paint to a color shows us that white paint lightens; adding black paint shows us that black paint darkens.

Follow-up
- Have the children describe the new colors they created.
- Compare dry mixtures with liquid mixtures.
- Discuss the children's observations when the paint was spooned onto the towel.
- Share with us the results of adding the white and black paint to make new colors.

Learning about dyeing and fading

Goal To discover how cloth absorbs color and how the sun fades colors.

New words Dye, fade

Teacher/Parent tips
- Children will need to wear painting smocks to protect their clothes and all table surfaces should be covered.

Lab needs
- 8-inch squares of white cotton cloth
- Food coloring
- Red, blue and green construction paper
- Small objects (e.g., rocks, leaf, twig, shell)
- Two separate pieces of brightly colored cloth
- Scissors
- A new and a faded T-shirt of the same color

Scientific principle
- The color of white cloth can be changed by dyeing.
- Strong light from the sun can fade colors.

Purpose The purpose of the following activities is to show how the color of a piece of white cloth can be changed by dyeing it a new color, and that the sun will fade bright colors. The experiments will also show that the hot sun can change a solid crayon into a liquid.

Children will discover the white cloth absorbs the colors that are squeezed on it. In contrast, they will discover that the sun fades both paper and cloth. This will show them that the strong light from the sun can fade our clothing.

Activities
1. Fold the 8-inch square of white cloth twice. Squeeze the food coloring onto the folded cloth in the design of your choice. Unfold and let dry.
2. Lay the sheets of brightly colored construction paper in the hot sun. Place small objects, such as a rock, leaf, twig, and shell, on the paper. Check under the object about every hour.

3. Cut each piece of brightly colored cloth in half. Place one half of each piece in the hot sun for a day or two. Observe what happens to the pieces of cloth and then compare them to the other halves that have not been in the sun.

4. Examine the two T-shirts. Decide which one was new and which one was faded.

Explanation In the first experiment, we saw we were able to dye the white cloth by adding color.

When we removed the objects from the colored construction paper in activity 2, the paper was darker where the objects had been. By comparing the paper left outside with a sheet still in the package, we can see how much the paper faded. This shows us that the bright rays from the sun are strong enough to fade (remove color) from the paper. Similarly, we learned that by putting the brightly colored cloth in the hot sun, it faded just like the paper.

As for the T-shirts, it was easy to see which was new and which was faded because the colors in the old shirt were not as bright. The clothing we wear fades if we spend a lot of time in the hot sun, after it is laundered numerous times, or through normal wear-and-tear.

Follow-up

- Discuss how the cloth absorbed the colors, and what happened when the colors ran together.
- Compare the sheets of paper after removing the objects.
- Remove the pieces of cloth from the sun and compare them to the pieces that were not in the sun.
- Discuss the similarities and differences in the two T-shirts.

Learning about melting crayons

Goal   To discover the effects of heat and cold on crayons.

New word   Rainbow

Teacher/Parent tips
- Cover work area with several layers of newspaper.
- Remove paper from crayons.
- Strings can be attached to the grated crayon projects and then hung in a sunny window.

Lab needs
- Wax paper
- Assorted broken crayons
- Hand-held pencil sharpeners
- Scissors
- Aluminum foil

Scientific principle
- Heat can be used to change some solids to liquid.

Purpose   The purpose of the following activities is to show how heat changes a solid to a liquid, and then by cooling the liquid, it will change back into a solid.

Children will discover what happens when crayons are left in the hot sun and how they can use this concept to make a grated crayon rainbow.

As they draw on the hot aluminum foil with a crayon, they will discover that the foil has absorbed enough heat to melt the crayon.

Activities   1. Place one sheet of wax paper in the hot sun. Grate the crayons onto the paper. When the grated crayons start to melt, place a second sheet of paper on top and press gently with your hand. Move the paper to the shade or a cool place. When cool, trim the paper into the desired shape.

51

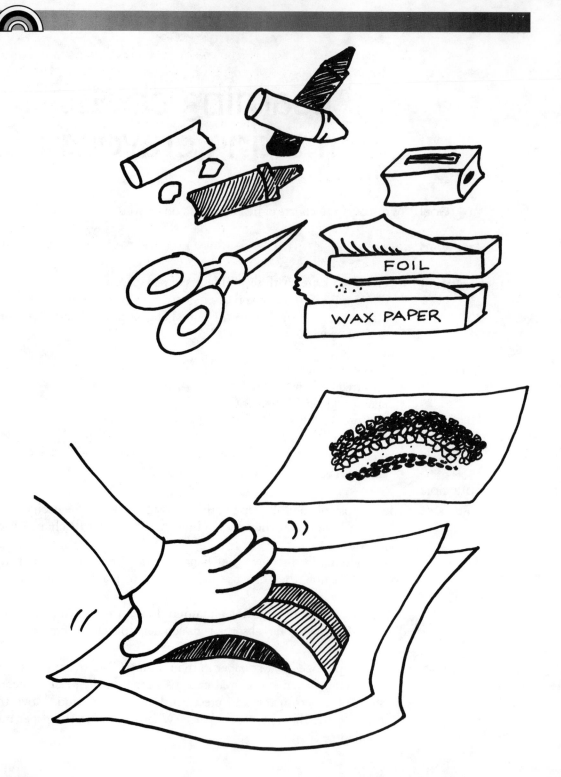

2. Place a sheet of wax paper in the hot sun. Make a rainbow by grating the colors onto the wax paper in arcs. Make the arcs close together in the following order: red, orange, yellow, green, blue, indigo, violet. When the grated crayon starts to melt, place the second sheet of paper on top, and press lightly. Move your rainbow to a shaded area or cool place. When cool, trim the edges of the paper.

3. Place a sheet of foil on a hot sidewalk. When the foil becomes hot, draw a design on it with crayons.

Explanation With our melted crayon projects, we discovered that the heat from the sun causes our solid crayons to change into liquid crayons. Yet, by moving the papers to a cool place, we saw the crayons harden and become solid again. This is another example of a cycle.

When we made a rainbow, we watched the colors melt and run together. In some of your rainbows, it is hard to pick out all seven colors, just as it is hard to identify all the colors in a rainbow because the colors blend together.

Drawing on the hot aluminum foil melted our crayon enough to leave a design. This tells us that the foil became hot from being in the sun, and then the hot foil melted the crayon as we were drawing.

Follow-up
  • Discuss the effects of heat and cold on the crayons.
  • Explain how the seven colors of the rainbow always appear in the given order, but that the colors blend so we can't always identify each color.

53

Part 4
Kitchen Science–
Stir up Some
Science

Through the activities in this section, our kitchen scientists will experience measuring, following a recipe, and preparing and cooking good things to eat.

They will mix and stir ingredients in their activities. Through hands-on experience, they will see how a liquid and a solid substance can be combined, changing both during the process.

Physical experiences, such as grinding peanuts to make peanut butter, will show them how they can change one food into a new food.

In the popcorn activities, children will discover the little surprise inside the kernel that makes dry popcorn pop when heated.

Our gardening scientists will be involved in growing their own sprouts and making a sprout sandwich, as well as discovering what causes the new growth from seeds.

While working together to make vegetable soup, the children will make other discoveries, such as no two vegetables are alike, all have a covering on the outside, and some are a different color inside than outside. The taste-testing conducted in this section will introduce the children to new vegetables.

Learning about measuring and mixing

Goal To become aware of the importance of measuring correctly.

Teacher/Parent tips
- Provide materials for washing utensils. Encourage recycling and clean-up.
- Demonstrate how to measure correctly (e.g., using a butter knife to level off a spoon).

Lab needs
- Measuring spoons
- Powdered sugar
- Clean margarine tubs
- Milk
- Food coloring
- Graham crackers
- Plastic knives
- Soft margarine
- Honey

Purpose The purpose of the following activities is to involve children in mixing and stirring experiences and to stress the importance of following a recipe and measuring correctly.

They will also have the experience of mixing ingredients together, spreading, and observing the changes that take place.

Children will discover changes in appearance, texture, and taste, and at the end of the experience, will have something to eat.

Activities
1. To make icing, measure 1 tablespoon of powdered sugar into a margarine tub. Stir in one-fourth teaspoon of milk. Add two drops of food coloring. Mix and spread onto two graham crackers.
2. To make honey butter, measure 1 tablespoon each of margarine and honey. Mix well, then spread onto graham crackers.

Explanation   While making our icing, we mixed two ingredients together and made something new—the icing. The appearance of the sugar and the milk changed after we mixed them together. Both were white in color to start with, but we discovered that by adding food color, we could change the color. Did you notice that we had to spread the icing on the cracker carefully or the cracker would break?

We found a way to change the color and the taste of butter by adding the honey. Stirring is a way of mixing two ingredients together. Stirring made the honey butter smoother.

Some of us found that if we didn't follow the recipe correctly, our icing was either too thin when we added too much milk or too thick when we added too much sugar.

Follow-up   • Discuss what happens when we don't measure correctly.
• Discuss what happens when a liquid is mixed with a solid.
• Describe and compare the taste, smell, color, and texture of ingredients before they are mixed and after they are mixed.

Learning about peanuts and popcorn

Goal To learn how to make peanut butter and to discover why popcorn pops.

Teacher/Parent tips
- Hand food grinders can often be found at swap meets or borrowed from older citizens.
- Each grain of popcorn contains moisture that, when heated, changes into steam. The hard covering keeps the steam from escaping so it explodes with a "pop." The dried kernels contain no moisture and will not "pop."

Lab needs
- Peanuts in the shell
- Hand food grinder
- Bowls
- Knives
- Crackers
- Popcorn
- Popcorn popper

Purpose The purpose of the following activities is to show that by grinding peanuts, a new product can be made, and that heat changes popcorn kernels into an edible food.

Children will discover and observe the change in appearance, texture, and taste after the peanuts are ground, and in the popcorn kernels after they are popped. By experimenting, our kitchen scientists will discover that through the grinding process, they are able to change the peanuts to peanut butter, and will understand that popcorn pops only when it contains sufficient moisture.

Activities

1. Shell the peanuts. Grind a few at a time in the hand food grinder. Continue re-grinding until the peanut butter is creamy and can be spread on crackers.
2. Pour some popcorn on a paper towel and set it in the sun or in a hot, dry place to dry out. Make some popcorn to eat. Try to pop the sun-dried corn, observing carefully for any pops.

Explanation When we examined the shelled peanuts, they looked and felt different from the peanuts in the shell. Some peanuts contained more nuts than others. On the first grinding, the peanuts came out of the grinder dry and crumbly, but the more times we put the ground peanuts back through the grinder, the smoother and creamer the mixture became. This is because peanuts contain moisture and oil. Finally we discovered a complete change in the peanuts—we had peanut butter!

Each kernel of popcorn has a tiny drop of moisture inside. This is what makes it pop. When the corn gets hot enough, the water pushes against the shell and causes the kernel to explode making a popping sound. The white part of the popped corn came from inside the kernel.

The kernels of corn we set in the sun dried out. Since no water was left inside the kernel, this corn did not explode when heated. It just got hot and started to burn. This shows us that popcorn that has dried out will not pop.

Follow-up • Have children explain the process of making peanut butter. Discuss the children's observations as the peanuts were being ground, as well as its smell, taste, and texture. Do the same for the popcorn making process.

- Compare whole peanuts to the peanut butter and corn kernels to the popped corn.
- Challenge children to explain why the dry corn did not pop. Explain the importance of moisture in the kernel.

Learning about edible seeds and sprouts

Goal To discover that seeds will sprout with or without soil.

New word Sprout

Teacher/Parent tips
- Edible seeds can be used for concept development, such as counting, sorting, discriminating by shape and color, and comparing similarities and differences through both visual observations and taste.

Lab needs
- Baby food jar
- Bean, lentil, cress, or alfalfa seeds
- Squares of cheese cloth
- Rubber bands
- Water
- Plastic knife
- Bread slices cut in quarters
- Soft margarine
- Soil
- Window boxes or pots
- Water

Scientific principle
- In order to sprout, seeds need warmth and water.

Purpose The purpose of the following activities is to show that seeds will sprout both with or without soil if they receive sufficient warmth and water.

    Children will discover that they can grow edible sprouts from some kinds of seeds. They will be able to watch the seeds sprout and grow, and then have a chance to make a sprout sandwich.

    By planting some of the same seeds in soil, they will see the seeds sprout as the new plant pushes up through the soil.

Activities 1. To grow sprouts, first cover the bottom of a baby food jar with a layer of seeds (e.g., bean, lentil, cress, alfalfa). Cover the seeds with water.

Cover the jar with cheese cloth and secure with a rubber band. Set the jar in a dark place overnight. Each day, rinse the seeds by allowing water to run through the cheese cloth, pour off the water, and return the seeds to the dark until green appears. At the first sign of sprouts, place the jar in the sun. Sprouts are ready to eat when they grow to 1 to 2 inches long.

2. Spread the pieces of bread with soft margarine. Sprinkle sprouts on top for a "sprout sandwich."
3. Plant some of the seeds in soil in window boxes or pots. Place in a sunny window and water regularly. Sprouts should appear in three to five days.

Explanation When we started our sprout jar, we first soaked the seeds overnight. As the seeds absorbed water, they swelled up and split the seed covering. By

keeping the wet seeds in a warm place, the sprouts kept growing. This shows us that in order to sprout and grow, seeds need water and warmth.

Before we sprinkled our sprouts on the piece of bread, we first buttered the bread to help keep the sprouts from falling off. The butter also gave our sandwich a good taste. We tasted several kinds of sprouts and each had a special taste.

Our sprouts are new plants that grew from seeds. The seeds we planted in garden soil also need warmth and water in order to sprout. These sprouts can also be eaten if we wash them well.

Sprouts from some seeds should not be eaten, so remember to eat only those sprouts that are grown for eating.

Follow-up
+ Compare the sprouting time of the seeds in activity 1 with those in activity 3.
+ Have the children describe the taste of the different sprouts.
+ Take a survey to determine which sprouts the children liked best.
+ Provide materials so that the children can draw pictures of the daily growth of the plants in activity 3. Use a different color to represent each day's growth.

Learning about vegetables

Goal To explore the attributes of a variety of vegetables.

Teacher/Parent tips • The edible part of a vegetable may be the bulb, flowers, leaves, roots, or stems.

Lab needs • A variety of vegetables
• Water
• Plastic, serrated knives
• Cooking pan and stove
• Spoons
• Bowls
• Mixing spoon

Purpose The purpose of the following activities is to become more familiar with the different types of vegetables, and to discover which part of the plant is eaten.

Children will discover similarities and differences in vegetables. They will be shown different ways to classify vegetables.

Activities

1. If in a classroom situation, have each child bring in one vegetable or take a trip to the grocery market and have each child choose one vegetable. Lay the vegetables on a table. Count them, then sort them by color, size, and texture.
2. Wash the vegetables, pare (only if necessary), cut into bite-size pieces, and place in a cooking pan. Add enough water to cover the vegetables and cook until tender but still crisp. Serve the vegetable soup.

Explanation In sorting our vegetables, we discovered that some were the same color, that a few felt rough on the outside, and that each had either leaves, roots, or stems. This shows us even though a vegetable may be the same color and feel the same, each one is different.

We cooked our vegetables to make a soup; however, most vegetables can be eaten raw. We found that when we peeled some vegetables, they were not the same color on the inside as on the outside. The peel protects the inside part of the vegetable just like our skin protects the inside part of us.

With our cooking experiment, we saw that vegetables make good soup. Some vegetables may taste a little like another kind of vegetable, but each one has a different taste and name. No two are exactly alike.

Follow-up
- Take turns describing the taste of the soup.
- Take a vote on the favorite vegetable in your soup.
- As you name each vegetable, have the children give input as to where it grows (e.g., above or below ground).

69

Part 5
Temporary Pets–Pet Sitters and Caregivers

This section introduces little biologists to small animals that make good pets for the classroom, and about the habits and needs of these animals.

The children will also learn to observe small animals (kept on a temporary basis) and perhaps have a chance to view firsthand their life cycles.

By becoming involved in caring for pets and other small creatures, children will see how in such situations animals are totally dependent on them for food, water, and care.

Learning about the habits & care of classroom pets

Goal To determine which animals make good classroom pets and how to care for them.

Teacher/Parent tips
- Borrow pets from parents or friends for a few hours or for the day. Ask them to write a short letter about the pet and to provide food and a cage, if necessary.
- White rats, hamsters, rabbits, and chickens are good all-day pets. Dogs, cats, and birds are good short-term choices.

Lab needs
- Pets on loan (one per day)
- Stuffed toy animals
- Pinecones
- Dried bread
- Birdseed
- Containers for water

Purpose The purpose of the following activities is to learn more about small animals, their needs, and how to care for them.

Children will discover that each borrowed pet has special needs as they read and follow the directions for its care that the owner provided.

They will become more aware of the birds around the play yard, how each type of bird is different, and that birds also require food and water.

Activities
1. Have children discuss the chores involved in taking care of their pets.
2. Examine the borrowed pet and read the letter that came with it from its owner. Follow the directions for its care.
3. Have each child choose a stuffed toy animal and tell how they would care for it. Use the stuffed animals to create a "pet farm" for free play.
4. Discuss with the children what food is safe to offer to wild birds. Stuff pinecones with pieces of dried bread. Place the cones, birdseed, and pans of water in the yard where birds are frequently seen. Make a color graph to show the different colored birds that visited.

Explanation   As we read the letters about caring for our visiting pets, we saw that each one had special needs. For example, our cat needed food only once a day, but we could keep food in the rabbit's feeder all the time. Animals are like us in that they need food and water. When they are kept as pets, they need someone to care for them.

In the play yard, we observed many types of birds. We watched them come for the food and water we set out for them. All animals need food and water, but each animal eats different types of food. For example, the birds ate our dry bread, but they did not eat the pinecone.

Our color graphs showed us that more brown birds came to visit us than any other color.

Follow-up   • Discuss the need to keep cages, food dishes, and water containers clean, and what was learned about the visiting pets.
   • Talk about the different types of birds that visited your playground or yard (e.g., which birds were the smallest and largest, how many different colored birds were observed).

Learning about jelly-jar pets

Goal   To learn about small animals by keeping them as temporary pets.

New words   Environment, habitat, cocoon, chrysalis, larva

Teacher/Parent tips
- Observation and discussion is the best way to learn about these "pets."
- They can be kept in jars for two days without food or water. After observing, return them to their natural environment.

Lab needs
- Wide-mouth, unbreakable jars with tight-fitting lids, punched with small holes
- Jelly-jar pets (e.g., tadpoles, small frogs and lizards, caterpillars, cocoons)
- Algae, leaves, pebbles, and twigs

Purpose   The purpose of the following activities is to help children understand more about some of the small creatures that live in and around the school or home. Children will also have an opportunity to see the changes within the life cycle of a frog, butterfly, or moth.

Children will discover, by keeping small animals in a jelly-jar for a short period, that they will be able to make close-up observations and comparisons relating to the animal's size, color, and texture. By providing food, water, and air, young biologists will come to appreciate a small animal's life.

Activities
1. Collect a jar of tadpoles, along with some algae and leaves, from a pool or stream. Since tadpoles change continuously, they are good pets to observe. Have the children draw pictures of a tadpole over a five-day period.
2. Small frogs, lizards, and horned toads can be kept in a large jar for a day. Place a few fresh leaves, pebbles, and a twig in the jar. Lay the

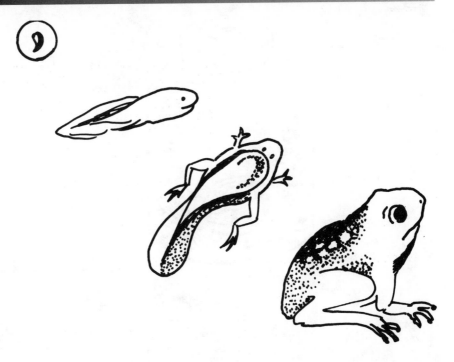

jar on its side. Observe their movements and other habits. Identify their similarities and differences.

3. Search the yard for caterpillars, cocoons, and chrysalises and place several in a jar. Collect plenty of leaves along with the caterpillars. Ask for input as to what might live in a cocoon or chrysalis. Compare the size, shape, and color of the collection.

Explanation An animal's natural home or where it usually lives is called its habitat. Each animal has its own special habitat.

We saw in the pictures of the frog cycle how eggs hatch into tadpoles and how the tadpole then changes into a frog. Our tadpoles looked like long fish when we first saw them. The next day, we saw that they grew two back legs, then two front legs. Finally it grew into a frog with a tiny tail, then a full-fledged adult frog. After our tadpole grew legs, we needed to put a rock in the water so when it became a frog it could hop up out of the water.

Frogs, lizards, and horned toads are helpful animals because they eat flies and other insects. Frogs live near the water, but lizards and horned toads like a warm, dry environment.

If we planned to keep the temporary pets several days in the jars, we

would have to provide them with food and water. If we put them back into their natural habitat, they will be able to get their own food and water.

In our collection of cocoons, we found large and small ones, glossy and fuzzy ones, and black and brown ones. We found different types of butterflies and moths when they broke out of their cocoons. We learned about the life cycle of the butterfly as we watched its changes—first the egg, then the larva or worm, next the cocoon or chrysalis, and finally the butterfly or moth.

Follow-up
- An encyclopedia contains good pictures of the life cycle of the frog and butterfly. These pictures will help you identify caterpillars and the inhabitants of cocoons.
- Discuss the life cycle of the frog, what butterflies eat, and how lizards aid the environment.
- Explain that frogs, lizards, and horned toads need a place to hide, that moisture from the leaves will provide all the water they need, and that we should keep these animals for only a short period to observe them, then they should be returned to their natural habitat.

Part 6
Ecology-Conscious Kids–Life in Our World

This section will encourage the young earth caregiver to recycle by introducing a number of ways to become involved. Children will discover that they can chop up a decayed pumpkin and collect the peeling from the oranges they ate and put them into the ground to make their own compost and to enrich the soil. Later they will be able to use the new soil for planting seeds.

Youngsters will have the experience of creating something new from something that is usually thrown away. Creating art from garbage and sculptures from boxes will provide another opportunity for them to become involved in recycling.

Learning about being a good earth caregiver

Goal To help children learn to take responsibility for caring for our world.

New words Litter, caregiver, recycle, ecology

Teacher/Parent tips
- *Ecology* is the interrelationship of organisms and their environments.
- Remind children to wash their hands with soap and water after collecting litter.

Lab needs
- Used paper and plastic bags
- Newspapers
- Poster board
- Felt marking pens

Purpose The purpose of the following activities is to help young children learn how they can aid the environment by becoming actively involved in litter programs and recycling.

Children will discover how they can help reduce litter and keep areas free of trash. They will learn about sorting materials for recycling, and will discover how careless and unconcerned some people are about our environment.

Activities
1. Use old bags to pick up trash around the school, home, or play yard. Keep recyclable items such as cans and bottles in separate bags. Tally the number of bags picked up over a period of a week.
2. Use old bags to pick up trash on a walk around the block. Keep recyclable items such as cans and bottles in separate bags. Tally the number of bags picked up over a period of a week.
3. Decide on an area around the school, home, or play yard that children will be responsible for keeping free of litter for a month or longer. Make and post signs similar to the ones seen on highways (litter removal by ...).

Explanation As we picked up the trash around the school, home, and play yard, we saw just how much litter there was in one small area of our world. We found a way to sort the litter so that we had items we could recycle. We also found litter on our walk around the block, which shows us many people in our community litter our environment.

We all became good earth caregivers as we picked up litter and recycled it, took on the responsibility of keeping the area around our school and home free of litter, and told others about our project.

Follow-up • Ask children to explain the terms ecology, environment, recycle, litter, and an earth caregiver.
  • Discuss how some of the litter could be recycled. If possible, share the projects with other classes or neighbors.
  • Compare the number of bags of trash picked up on the school or home area to the number of bags collected on the walk.

Learning about compost

Goal　To experience the process involved in creating compost and its value.

New word　Compost

Teacher/Parent tips
- Compost consists mainly of decayed organic matter.
- Serve oranges for a taste experiment and save the peelings.

Lab needs
- Old jack-o'-lantern
- Orange peelings
- Pumpkin and bean seeds
- Plastic knives
- Spoons or small shovels

Scientific principle
- Over a period of time, garbage that is buried will decompose.

Purpose　The purpose of the following activities is to help children understand that by burying plant garbage and waste, we are both recycling and enriching the soil.

　　Children will discover that plant waste buried will in time decompose and mix with the soil.

Activities

1. Observe a pumpkin as it decays. Chop it into small pieces and bury it under about 4 inches of soil. Water well. Mark the spot.
2. Cut orange peelings into small pieces and bury them under about 4 inches of soil. Water well. Mark the spot.
3. Periodically predict what is happening to the pumpkin and orange peelings. After a month, dig into the ground where the peelings were buried. If the fruit is not decomposed, cover it, water it and wait another week. When decomposed, mix the compost into the surrounding soil and plant the seeds. Keep the soil moist.

Explanation When we put our old pumpkin and the orange peelings into the ground, we reduced the amount of garbage we threw into the trash can. We made compost by mixing our garbage with the soil. The compost mixed into the soil makes it richer and lighter so more air can enter. This helps new plants grow faster and stronger.

We examined our new soil, we found it was darker in color, felt lighter, and was easier to dig than the soil that had no compost.

We help the environment when we bury our biodegradable garbage into the soil so it does not have to be hauled to a trash collection place. We are good earth caregivers because we take good care of our earth.

Follow-up • Explain how organic matter enriches the soil.
• Discuss what we thought the buried material would look like in relation to the actual results.
• Encourage input as to why composting aids the environment.

Learning about garbage art

Goal To make children aware that some types of "garbage" can be reused for creating art projects.

Teacher/Parent tips
- Fruits and vegetables can be used for taste experiments and the tops, peelings, and seeds saved for the projects.

Lab needs
- Textured peelings from fruits and vegetables
- Seeds
- Tops cut from root vegetables
- Leaves
- Layers of paper towels saturated with liquid paint
- Ink pads
- Paper
- Glue
- Wax paper
- An iron

Purpose The purpose is to introduce children to another way of recycling by using the nonedible parts of fruits and vegetables for art projects.

Children will discover artistic ways in which peelings, leaves, and other discarded parts of fruits and vegetables can be used. They will be able to express their creativity in new and interesting ways through the use of these new printing, stamping, and designing "tools."

Activities

1. Use various parts of fruits and vegetables as stamps. Create stamp-pad printings.
2. Glue the same type fruit and vegetable parts used in activity 1.
3. Place leaves between two layers of wax paper and press with a warm iron. Let cool.

Explanation We found we could recycle parts of fruits and vegetables, which we usually throw away, to make interesting prints and designs. We used seeds, leaves, and peelings as printing tools and found that each one made a different stamp-pad design. We were also able to match the plant parts to the designs we made.

The warm iron, pressed across the wax paper, sealed our leaves between the two pieces of wax paper, but did not burn the leaves.

Follow-up
• Compare the various types of art projects that were conducted, and identify the material used in each stamp print.
• Discuss other uses for the "garbage."

Learning about recycling throwaways into craft projects

Goal To encourage children to recycle throw-away items before they are discarded.

Teacher/Parent tips • A few weeks before doing the following activities, have children collect small cardboard boxes, scraps of fabric, yarn, and other decorating materials.

Lab needs (Material to be collected by children)
• Small cardboard boxes
• Scraps of fabric
• Yarn and other decorating materials
• Glue
• Scissors
• Large pieces of corrugated cardboard
• Paper bags

Purpose The purpose of the following activities is to help young scientists develop the habit of saving and recycling, and to think of ways to use throw-away items for art projects.

Children will discover that items that we often discard can be transformed into innovative art projects, toys, and play units.

Activities
1. Sort the collected recyclable items(e.g., small cardboard boxes, scraps of fabric, yarn, other decorative materials). Discuss possibilities for recycling the materials (e.g., box sculptures, puppets, animals, buildings). Make a box sculpture.
2. Using the small boxes, work in groups to create a farm and farm animals, a city street and buildings, and a transportation center. Glue the newly created scenes to large pieces of cardboard.
3. Decorate a box and its lid to be used as a container.
4. Bag all leftover scraps for future use.

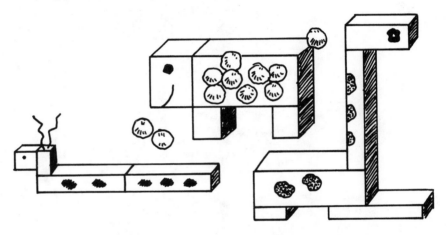

Explanation We collected many items for our projects and by recycling, we created many new things. This shows us that we should never throw anything away without first trying to find another use for it.

By sharing your projects and bags of leftovers with a friend, we taught others ways to recycle and to become good earth caregivers.

Follow-up
• Discuss the art projects that were made and then make a picture chart or graph to show how many new items were made from the recycled items.
• Use the projects in activity 2 for creative play.
• Send home a bag of leftover materials with each child. Suggest that they share one of the classroom projects with a friend or family member.

Part 7
Air–It's All Around Us

Through the experiments and activities in this section, the little scientists will discover that even though we cannot see air, it is all around us, and that wind is the movement of air.

They will learn about moisture in the air and about materials that absorb and repel water. The little scientists will experience the process of evaporation by experimenting with a variety of materials.

The fact that air is all around us, it can be felt, and it has weight are just a few of the scientific principles relating to air that are explored.

Youngsters will discover that wind, though not always welcome, has many helpful uses and that we depend on the wind for moving seeds, soil, and the clouds.

Learning about air

Goal To discover some of the qualities of air.

New words Air, wind

Teacher/Parent tips
- Air is a mixture of gases that contain water in the form of vapor.
- Air takes up space and has weight.
- Warm air rises.

Lab needs
- A soft, clear-plastic bottle
- Yardstick
- Two balloons of equal size
- One long piece of string and two equal lengths of string
- An easy-to-read thermometer
- Yardstick
- Tape

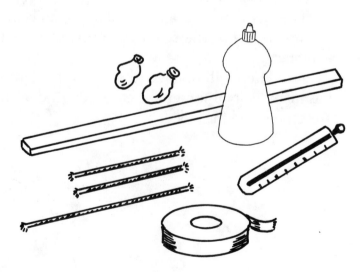

Scientific principle

• Air takes up space and has weight.

Purpose

The purpose of the following activities is to give young children an understanding of what air is, how to describe it, and why it exists.

Children will discover that although a container looks empty, it contains air. The activities show that even though we cannot see air, it takes up space and has weight.

The little scientists will experience the movement of air as they inhale and exhale. Furthermore, they will be able to prove that warm air rises.

Activities

1. Have the children give their answers to the following riddle:

   > What is all around us,
   > takes up space, has weight,
   > but cannot be seen?

2. Display a soft, clear-plastic bottle and ask the children if they think it is empty. Take turns squeezing the bottle with the open end against your face.

3. Tie a long piece of string around the center of a yardstick. Take two strings of equal length and tie one around the top of a deflated balloon and one around and an inflated balloon. Make a loop at the end of both of these strings. Hold the ruler by the long string and explain to the children that it is balanced. Have a child slip the loops from both balloon strings over each end of the ruler. Discuss what happened.

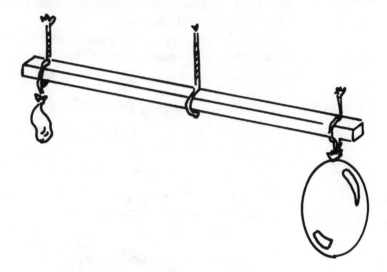

4. Place your hand in front of your face. Take a deep breath and exhale first through your nose and then through your mouth.
5. Tape an easy-to-read thermometer to the end of a yardstick. Place it on the floor for about three minutes, then record the temperature. Hold the same yardstick up near the ceiling for about three minutes, then record that temperature.

Explanation Even though we could not see anything in the bottle, we discovered that there was air inside when we squeezed the bottle. We cannot see air but we can feel it.

We blew air into the one balloon and found it was heavier than the balloon that was not blown up. This shows us that air has weight.

By exhaling or breathing out, we were able to feel the movement of air, which shows us we breathe new air into our body and breathe out the old air. We need air to keep us alive.

The temperatures taken with the thermometer showed that it was warmer closer to the ceiling than it was on the floor. This proved that warm air rises.

Follow-up
- Discuss what activity 3 proves.
- Explain where the air in our breath comes from.
- Compare the temperature at floor level to the ceiling temperature.

Learning about moisture in the air

Goal   To learn about evaporation and absorption, and to discover that air contains moisture.

New words   Repel, evaporation, absorption

Teacher/Parent tips
- Review rain and cloud formations from Part 2.
- *Evaporation* is the drawing off of moisture in the form of vapor. *Absorption* is the ability of a material to suck up or take in.

Lab needs
- Large paintbrushes
- Cans of water
- Items that will absorb water
- Kitchen timer
- Pie tin
- Salt
- Hand mirror

Scientific principle

• Air contains moisture.

Purpose

The purpose of the following activities is to learn more about evaporation and absorption and to show that some materials repel water. The little scientists will also do experiments to prove that air contains moisture.

Children will discover that there is moisture in the air around us and in the air we breathe. Their understanding of water evaporation and absorption will be reinforced as they participate in the activities. They will also find that there are many materials that will not absorb water.

Activities

1. Paint the sidewalk with water.
2. Place wet objects that absorb water, such as a newspaper, a sponge, and a piece of cloth, in the hot sun. Set the timer and check the objects periodically.

3. Cover the bottom surface of a pie tin with salt. Set the pie tin outside overnight. After recording what happened overnight, place the pie tin in the hot sun and record your observations.
4. Open your mouth and exhale onto a mirror.
5. Have the children collect items that they think will not absorb water.

Explanation

We watched the water paint on the sidewalk disappear. This shows us that the heat from the sun caused the water to evaporate. We also found that heat from the sun would evaporate water from other materials that absorb water, such as the wet newspaper, the sponge, and piece of cloth. The water or moisture went into the air, proving that there is moisture in the air.

In activity 3, we found that the salt that was left outside overnight absorbed moisture from the air, and if set in the hot sun, the heat of the sun evaporates the moisture.

By breathing on the mirror we were able to see moisture on the mirror. We know we breathe air so this was another way we proved that there is moisture in air.

Follow-up
- Discuss what happened to the water on the sidewalk.
- Explain absorption and evaporation. Collect items from the classroom or your home and decide which ones will absorb water.
- Check the pan of salt early in the morning and talk about the changes in the salt.
- We can find out if some of the items you selected will repel water. Experiment by pouring water over the items that are safe to experiment with. Do the experiment outside or over a pan.
- Ask for an explanation as to why raincoats and boots keep us dry in the rain (e.g., they do not absorb water, but repel water).
- Discuss the discovery made when we exhaled onto the mirror. Discuss where else moisture in the air may come from.

Learning about wind

Goal To become aware of the helpful qualities of the wind.

New words Wind, energy

Teacher/Parent tips • Wind, a form of energy, is the movement of air.

Lab needs
• Sheets of newspaper
• Grass seed
• Paper plate

Purpose The purpose of the following activities is to help young scientists learn about wind and that it is a form of energy. Although they cannot see the wind, the activities will enable them to see the effects of wind.

Children will discover that wind is the natural movement of air, it causes things to move, and it has strength and force. They will be able to observe the effects of wind and how it aids nature by distributing seeds and moving soil.

Activities
1. On a windy day, hold an opened-up sheet of newspaper in your hands while standing in an open area. Then, tear two strips from the newspaper. Hold one in each hand and run into the wind.
2. Go for a walk on a windy day.
3. Stand in the wind while holding an inverted paper plate in your hand. Sprinkle grass seeds on the plate and see what happens.

Explanation With the newspaper activity, we were able to watch the wind blow the paper. This showed us that wind is air in motion and that wind has power—it moved the paper.

We felt the wind against our face when we walked into it. We also felt the strength or power of the wind as it pushed against us, making it harder to walk. When the strong wind was behind us, we felt it pushing us, making it easier to walk. We saw the wind moving dirt, dust, clouds, and old leaves off the trees.

We proved again that wind has power and is a form of energy in the third activity when it blew the seeds from the paper plate. This is the same way the wind blows seeds from one place to another in nature. The wind also blows dirt to cover the seeds so they can grow.

In learning about the wind, we saw that the wind is a form of energy. This means when the wind blows, it puts out power or strength and moves things.

Follow-up
- Explain that although we can't see wind, we can observe its action. Discuss the effects of the wind on the newspaper.
- Share examples of the wind at work on our walk, such as blowing dirt, leaves, and trees.
- Explain that wind transfers topsoil from one place to another, distributes seeds, rotates windmills, and moves the clouds about.
- Discuss times when we might be able to smell and taste something in the wind, such as smoke from a nearby campfire and salt from the ocean.

Part 8
Energy–The Force that Moves

Energy was introduced in the last section through experiences with wind. This part deals with learning about three forms of energy—nature, machine, and kid energy. Other activities on wind energy are also included to show the wind at "work."

In observing appliances that operate by battery and by electricity, children will see how they are involved in starting and stopping a flow of energy.

By watching large machinery in operation, the children will be exposed to another form of energy being produced. Comparing the forms of energy will help them understand the need for different sources and strengths of energy.

In the last group of activities, kids will produce their own energy. They will become aware of how body energy is used to cause our body parts to move. The little scientists will come to understand how their bodies produce energy to enable us to make things move.

Learning about nature's energy

Goal To learn that rain and wind are forms of energy.

New words Nature, erosion

Teacher/Parent tips • Energy is the force that causes something to move or to alter its direction. Nature, machines, and kids produce energy.

Lab needs
- Sprinkling cans
- Water
- Toy sailboat
- Tub of water
- 3-foot-long strip of crepe paper
- Sheets of newspaper
- Paper and plastic bags
- Different-sized cardboard boxes

Scientific principle • Without energy, nothing can move.

Purpose The purpose of the following activities is to show the force of rainwater and wind as forms of energy. Although these forces can be destructive, the little scientists will learn that nature's energy is needed.

Children will discover that water is a form of energy that destroys, but that we also depend on the energy that is produced by water. One of the activities demonstrates how water causes soil to erode.

Activities

1. To show the force of water as a form of energy, have children build castles, lakes, and dams in wet sand or dirt. Sprinkle the structures with water. Continue adding water until a good example of erosion occurs.
2. To show the force of wind as a form of energy, hold one end of a strip of crepe paper while standing in an area sheltered from the wind, in the wind at ground level, and in the wind at the top of a slide or climbing bars.

3. Set a toy sailboat in a tub of water and place the tub in the wind.
4. Place sheets of newspaper, paper and plastic bags, and different-sized cardboard boxes in the yard on a windy day. Observe the wind at "work." Be sure to retrieve all materials after the observations.

107

Explanation   We used our energy to build the sand castles. When we poured the water on the sand castles we saw how water washed or wore the sand away. We call this wearing away erosion. The faster we poured the water, the faster the sand eroded. This is like the rain. Gentle rain over a long period will slowly wash away at rocks, mountains, and soil; storms and heavy rain will speed up erosion. Wind also causes erosion.

We found that the energy from the wind made our strips of crepe paper move, but they would not move by themselves when there was no wind. If no wind was present, we had to use our energy to make them move. Nothing moves without energy.

We saw the energy produced by the wind at work as it blew light objects like sheets of newspaper, boxes, and other trash across and out of the yard. This is another reason not to litter.

In our experiments, we saw nature's energy—the wind—move our strips of crepe paper and sailboat for play, but it can also move large, heavy sailboats to provide transportation. Nature's energy—water—can be used to produce electricity.

Follow-up
- Explain how the water from the sprinkling cans acts the same as rain and floods to wash away soil (erosion).
- Have children share their experiences with their strips of crepe paper in the wind.
- Discuss how the wind had the power to move the sailboat and the objects placed in the yard.

Learning about machine energy

Goal To help children understand that energy can be produced by machines.

Teacher/Parent tips
- Any safe, small appliances can be used. An adult should be responsible for plugging appliances into outlets.

Lab needs
- Small hair dryer
- Flashlight
- Battery-operated toy
- Wind-up toy
- Pictures showing examples of different types of energy producers (e.g., windmill, truck, lawn mower, toaster)

Purpose The purpose of the following activities is to help the little scientists learn how electricity, batteries, and springs provide sources of energy. They will also learn that people must be involved in the procedure in order for these sources to make something move.

Children will discover that they are responsible for activating the source of energy that will make their wind-up or remote control toys work and start electric appliances.

By visiting a place where heavy equipment is being operated, children will have an opportunity to observe different ways in which machines use energy to perform specific tasks.

Activities

1. First, plug in a small hair dryer. Turn on its switch and feel the air that comes out. Listen to the sound of the hair dryer. Next, turn on a flashlight. Then, take it apart to show that the batteries produce the energy to light the bulb. Remove the batteries and try to turn on the flashlight again. Experiment with battery-operated toys.
2. Arrange to visit an area where machines are being operated, such as a gasoline lawn mower, street sweeper, and heavy equipment to observe different machines that produce energy. If it is impossible to arrange such visits, show the children pictures of different types of energy-producing machines and discuss them.

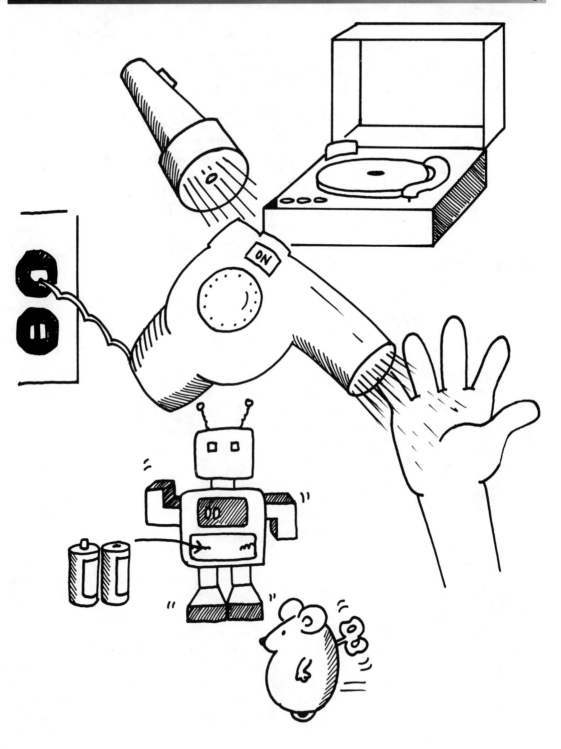

111

Explanation   We saw electrical energy make the hair dryer work, but it only worked when it was plugged into an outlet and turned on. The flashlight worked only when the batteries, its source of energy, were in place and it was turned on. The same thing happened with the battery operated toys, they did not work without the energy source.

A toy car can be moved by pushing it, using a wind-up spring, or utilizing a battery. The car cannot move by itself, it needs people either to push it or to start the energy working. Only nature's energy works on its own.

Follow-up   • Discuss the similarities and differences between battery-operated and electric energy producers.
    • Talk about the work done by the energy-producing machines.

Learning about kid energy

Goal   To experience ways in which physical energy can cause things to move.

Teacher/Parent tips
- Allow children time to investigate and experiment with the materials.
- Review nature and machine energy before going to the playground.

Lab needs
- A wind-up clock
- Kitchen timer
- A toy
- Egg beater
- Paper fan
- Strips of tissue paper
- Hand-operated tire pump
- Inner tube
- A wagon
- Kick ball
- Jump rope
- Nails
- Pieces of soft wood
- Hammers
- Shoe box lid
- Tennis ball

Purpose   The purpose of the following activities is to help children become aware of how they are an energy source that can cause things to move. They will see that by putting their energy into winding up toys or plugging in appliances they can make them work.

They will experience how the air in their lungs becomes a source of energy and how they can use their energy to operate a tire pump, which in turn produces energy to inflate an inner tube.

Children will discover, through activities and experiments, that they are energy producers and that there are some different ways they can use their energy. They will also discover how they control the expenditure of their energy to make things start and stop.

Activities

1. Experiment with a wind-up clock, a kitchen timer, a toy, an egg beater, and a paper fan to see how to make them work. Hold one end of a strip of tissue paper in front of your face and blow on it. Take turns pumping up an inner tube with a hand-operated tire pump.

2. Spend time on the playground swinging, bike riding, and pushing and pulling a wagon. Take turns kicking a ball to each other and jumping rope.

3. Work in small groups, take turns pounding nails into piece of soft wood with a hammer.

4. Place a shoe box lid on a small table and set a tennis ball inside the lid. Challenge children to find ways to make the ball move without touching it. Possibilities include tipping the lid, blowing on the ball, and lifting or pounding on the table.

Explanation In our first group of experiments, we used our energy to wind up a clock, a kitchen timer, and a toy, to beat eggs, to blow strips of paper, and to blow-up an inner tube. We put our energy into all of these to make them work.

On the playground, we used our energy to make the swing, bike, and wagon move, to kick a ball, and to jump rope. We found that it took energy to pound a nail into a board. Our energy made our muscles do what we wanted them to do.

We found several ways to make the ball in activity 4 move. But first we had to think before we could tell our muscles what to do. Pounding on the table produced the energy that was needed to move the ball, and by blowing on the ball we used energy to push the air quickly out of our lungs to move the ball.

People and animals get energy, not from batteries or electricity, but from food. Food gives us strength to move our muscles to bring about the movement of other objects, such as a swing or a bike. Our muscles also get energy from the food we eat. This shows us we need to eat healthy food and get regular sleep to keep up our energy.

We have learned about nature, machine, and kid energy. Some of the other sources of energy come from gas, which is used to heat our homes; from gasoline, which makes our cars move; and from the sun (solar energy).

Follow-up • Discuss how we can produce energy to cause things to move and to make objects or tools work for us.
• Share other ways we use energy that is produced by our muscles.

Part 9
Water–We Can't Live without It

The group of activities in this section will reinforce young children's understanding of water as they are introduced to new experiments using water, such as how a water drop acts as a magnifying glass and the effect of water on soil, sand, and gravel that causes erosion. The little scientist will learn about bubbles, how they are formed, and what holds them together.

In setting up a water-bottle aquarium, youngsters will learn to plan and work together to provide a balanced water environment where plant and animal life will be self-sufficient.

Learning about water

Goal   To discover some of the basic characteristics of water.

Teacher/Parent tips

- Show that water exists in three states—liquid, vapor, and solid—by displaying a glass of water, by boiling water in an electric coffee pot to produce steam, and by showing an ice cube.

Lab needs

- Kitchen baster
- Wax paper
- Newspaper
- Water
- Measuring cup filled with water
- Jar
- Bowl
- Flat pan

- Paper cup
- Three paper cups with an equal number of holes punched in the bottom
- Sand, soil, and gravel

Purpose   The purpose of the following activities is to reinforce the children's understanding of water and to learn more about it through additional experiments and observations.

Children will discover how a drop of water acts as a magnifying glass. Through another experiment, the little scientists will come to understand that the amount of water remains the same regardless of the size of the container it is poured into, and the amount of water cannot be increased through freezing.

Pouring water through different mediums will show how water soaks into different types of soil.

Activities   1. Examine water by squeezing several drops from the baster onto a piece of wax paper. Place small print from a piece of newspaper under a sheet of wax paper. Squeeze a drop of water over a letter. Squeeze a few drops of water onto the newspaper.

2. Note the water level on a measuring cup that is full of water. Pour the water into a jar, then a bowl, and then a flat pan. After each pouring, ask if the little scientists think there is more or less water in the container than there was in the measuring cup.

3. Mark the water level on a paper cup. Place it in the freezer for about an hour or overnight. Remove the cup of ice and check its water level. Set it in the sun.

4. Take three paper cups with an equal number of holes punched in the bottom and fill one with sand, one with soil, and one with gravel. Predict what will happen if we pour a cup of water over each. Pour the water over each material at the same time.

Explanation When we dropped the water on the wax paper, we saw it was not absorbed but formed a bubble or bead onto the wax paper. Wax paper repels water. By placing print from the newspaper underneath the wax paper, we saw that the print was enlarged as we looked through the bubble. The bubble acted like a magnifying glass. This did not happen when water was dropped onto the newspaper, for the newspaper absorbed the water.

By pouring the water back and forth into the three containers, we were able to see that the size or shape of the container did not change the amount of water. We had one cup of water when we started the experiment and after we finished experimenting, we still had just one cup of water.

In the next activity, we saw that the water expanded when it was frozen, but when it melted we still had the same amount of water. This shows us that although it looked like we had more water when it was frozen, we still had the same amount.

Our observations showed us that the water ran through the cup of gravel faster than through the cups filled with sand and soil, and it took the longest amount of time to run through the cup of soil. This proves that soil will hold water longer than sand or gravel.

Follow-up
- Share observations made when the water was dropped on the wax paper and then on the newspaper.
- Let children freely experiment with the cup of water and three containers until they discover that the amount of the water remains the same even when poured into different containers (based on Piaget's theory of conservation).
- Explain how the experiment in activity 3 shows that water expands when frozen.
- Compare the predictions made before conducting activity 4 to the outcome, and discuss the observations.

Learning about bubbles

Goal To discover bubbles and how to make them.

New word Bubble

Teacher/Parent tips
- Review how "kid energy" works.
- A black, plastic table covering makes it easier to see the bubbles.
- A *bubble* is a thin film of liquid that is inflated with air.

Lab needs
- Soft, clear-plastic bottle
- Tub of water
- Thick, liquid dish soap
- Paper cups
- Water
- Measuring tablespoon
- Drinking straws
- Black plastic table covering (e.g., plastic garbage bags)

Purpose The purpose of the following activities is to give the little scientists an understanding of how bubbles can be made, that bubbles contain air, and that how surface tension holds the air inside the bubble.

Children will discover that air from their lungs can inflate bubbles, and that an empty bottle contains air and the air in the bottle can be used to inflate bubbles. They will experiment with ways to create bubbles and will find out why bubbles break.

Activities
1. Squeeze a soft, clear-plastic bottle rapidly while holding it under water that contains thick liquid dish soap. Allow the children to create bubbles by churning the water with their hands. Squeeze the bottle under water again.
2. Partially fill cups with water and take them outdoors. Add a table-

spoon of liquid soap to each. Blow through a drinking straw into the cup forming bubbles. CAUTION: Tell children not to suck on the straws.

3. Cover an outdoor table, set in a sheltered area, with a black plastic table covering (e.g., plastic garbage bag). Fill a plastic bottle with equal amounts of liquid soap and water. Mix gently. Pour a large circle of bubble mixture in the center of the table. Give each child a straw. Insert the straws into the mixture and blow gently.

Explanation A bubble is a thin film of liquid that is filled with air. We made bubbles by squeezing the empty bottle under the water because there was air in the empty bottle. We pushed the air out of the bottle into the water to make a bubble. We found we could also push air into soapy water bubbles by stirring up the water with our hands. We made more and larger bubbles with the bottle after the soap was added. This is because the soap helps hold the bubbles together.

By blowing through the straw into the cup of soapy water, we used the air from our lungs to inflate the bubbles. The more air we blew, the more bubbles we made.

When we worked together to blow a giant bubble in activity 3, we discovered that if we blew gently, the bubble would keep growing larger and the thin film around the bubble mixture would hold the bubble together. The outside of the bubble acts like an elastic skin to hold it together. The more air we blow into the bubble the more the skin stretches. When someone blew too hard, it broke the skin and let the air out, so we no longer had a bubble. The black plastic on the table made it easier for us to see the bubble.

Follow-up
- Discuss how the bubbles were made. Compare the bubbles made without the soap to those made after adding the soap.
- Discuss what happened to the liquid in the cup after air was blown into it.
- Compare the bubble made on the table to the glass in a magnifying lens. Discuss what happens when we blow the bubble mixture too hard.

Learning about setting up a water-bottle aquarium

Goal To learn how to set up a balanced environment for small fish.

New word Aquarium

Teacher/Parent tips
- Parents or a bottled-water company may provide you with a clean, clear five-gallon bottle.
- If the plant and animal life is well balanced, the aquarium will require little care and will provide years of enjoyment.

Lab needs
- Five-gallon, clear-glass or plastic water bottle
- Marbles, small shells, and pebbles
- Water
- Liquid soap
- Three water snails
- Aquarium plants
- Three goldfish
- Three guppies

Purpose The purpose of the following activity is to help young scientists learn more about small fish, water snails, and aquarium plants as they work together to make an aquarium from a recycled water bottle.

Children will discover the joy of creating a special environment for water animals and plants. They will make many other discoveries as they become involved in setting up their aquarium, selecting the fish, and then observing how the animal and plant life are dependent upon one another. The little scientists will be able to observe the steps in another life cycle.

Activities
1. Plan how you are going to use the five-gallon, clear water bottle to set up an aquarium. List the items you will need on a picture chart.
2. Wash the marbles, shells, and pebbles with liquid soap and rinse them several times. Place these materials on the bottom of the clean bottle. Fill the bottle with water to a level right below the neck of

the bottle. Let the water set for 24 hours. Place the aquarium in good light but not direct sun.

3. Make arrangements to visit a pet store and for the owner to talk to the children about aquariums and fish. Have the children observe the fish in the store. Decide on the ones to purchase (they will often give you the snails and aquarium plants free).

4. Place the snails and aquarium plants in the bottle. Then add the fish.

Explanation On our visit to the pet store, we saw many types of fish, but we had already made a list of what we had planned for our special aquarium. This meant we had to buy the right number of fish and snails.

It was important to make sure the bottle, marbles, shells and pebbles were clean so we would not put anything into the aquarium that might carry harmful germs.

We put the fish in last so we would have their new home ready for them.

The space at the top of the bottle allows for plenty of air to enter the bottle. By choosing a clear bottle, we are able to see the fish swim up to get air. The bottle also lets us observe the fish's life cycle. That is, the fish eats the aquarium plants, then passes off waste which helps the plants to grow. Some of the waste from the fish is also eaten by the snails, which keep the grass from growing too fast and filling up the bottle. As long as this keeps working, the fish, snails, and grass will live. This means we have a balanced aquarium.

Fish that live in lakes and oceans also live in a balanced environment much like our aquarium. By using the water bottle to make our aquarium, we are recycling the bottle.

Follow-up • Discuss the need to let the water set out for a 24-hour period to allow for evaporation of chlorine. Also discuss the need to keep the water at the same level so that sufficient air will be allowed to enter to oxygenate the water for the fish.

• Observe the fish coming to the surface for air and how the grass provides both food and shelter. Explain that once the plant and animal life are balanced, little care is needed. The snails act as a clean-up squad! However, you may wish to add a small amount of fish food once a week and occasionally use a clean, wooden dowel to stir up the bottom.

Part 10
Tiny Creatures–The World Is Full of Them

This section will introduce children to plants and animals that live underground, to three of nature's tiny scavengers, and to a few-easy-to-find-and-observe insects as well as spiders.

By marking off small grassy areas in the yard to examine, the little scientists will be able to make a number of interesting discoveries between the blades of grass and around its roots. As they dig in their soil "discovery spaces," they will be surprised at what they may find living and moving about underground.

Snails, slugs, and pill bugs make fascinating subjects to observe. Young explorers will see how these tiny creatures aid the environment by acting as a "clean-up squad" much like the water snails in the aquarium we made.

Insects and spiders will be another area to investigate. Children will learn how to identify an insect, then compare the habits and appearances of insects and spiders.

Learning about what lives underground

Goal To makes discoveries about life underground.

New word Aphids

Teacher/Parent tips
- Locate (or prepare) a damp, soft soil area (where earthworms can be found) and a grass area.
- Locate a trail of small ants.

Lab needs
- Hula hoops
- Large spoons for digging
- Yarn
- Twig
- Sugar
- A leaf or flower bud containing aphids (i.e., small insects that eat plant juice)

Purpose The purpose of the following activities is to help children understand that there is life underground. By digging into the ground and between blades of grass, they can observe firsthand underground life.

Children will discover tiny creatures, insect eggs, and roots from trees and plants that live and grow below the ground.

By observing a trail of ants, they will see that, although ants live underground, they spend a good deal of time above ground searching for food and water.

Activities
1. Use hula hoops to mark off "discovery spaces" in damp soil. Have the children dig with a spoon in the soil within the hula hoop area to see what they can find.
2. Use hula hoops to mark off "discovery spaces" in a grassy area. Have the children use a spoon to separate clumps and blades of grass to see what they can find.

3. Observe a trail of ants. Try to find out where they are coming from and where they are going. Use yarn to measure the length of their trail. Cut the yarn and hang the "yarn trail" in the classroom. Place a small twig across the ant trail, wait a few seconds, then remove it. Sprinkle a few grains of sugar on their trail. Place the aphid-covered flower bud across the path.

Explanation As we dug in the ground within our hula hoop spaces, we found a variety of living things. Some types of worms make their cocoons underground, snails and some insects lay their eggs just beneath the ground, and earthworms move in and out through soft soil. We also found roots of trees and other plants, flower bulbs, and sprouting seeds underground.

In looking at the grass in our hula hoop spaces, we found that the part of a blade of green grass close to the ground was almost white. This is because it doesn't get the sun. Some small bugs and spiders look for food among the grass. They can also get a drink from the moisture that collects between the blades of grass.

We found a long trail of ants and followed their trail to their underground home. Most ants live in dens with many tunnels under the ground. Although they live underground, they spend a lot of time during the day above ground looking for food to carry back and store in one of their tunnels. Because ants like sweet things to eat, they stopped where

133

we sprinkled the sugar and carried the grains of sugar back to their den. They didn't eat the flower or the tiny aphids; however, they ate the sticky juice that covers the aphids.

Follow-up
- Discuss what was found in the soil. Explain that earthworms eat dead plant material in the soil. The holes they make as they push through the soil allows air, necessary for plant growth, to enter the soil.
- Examine the body of a worm with a magnifying glass. Note that the body is made up of segments. Note also that they have no eyes or ears, but they do have a mouth.

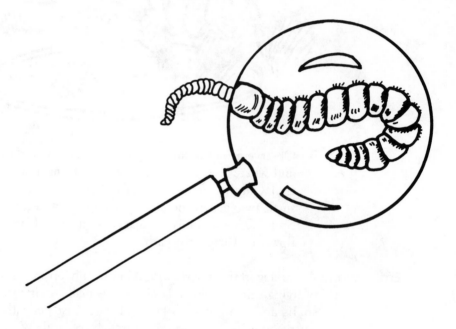

- Compare the color of the grass above and below the ground.
- Talk about the roots, weeds, and insects that were discovered.
- Share observations about the ants and their trail, and what discoveries were made when objects were placed in their path.

Learning about nature's scavengers

Goal   To discover the role that three tiny animals play in the balance of nature.

New words   Scavenger, vivarium

Teacher/Parent tips
- Note the nearby location of snails, slugs, and pill bugs.

Lab needs
- Magnifying glasses
- Damp soil
- Plastic shoe box
- Cheesecloth
- Tape

Scientific principle
- All living creatures contribute to the environment.

Purpose   The purpose of the following activities is to provide an opportunity for children to observe three of nature's scavengers—snails, slugs, and pill bugs—both in their natural environment and in a classroom environment. The activities will also help them understand how these tiny creatures contribute to our environment.

Children will discover the similarities and differences between snails and slugs, where they lay their eggs, what they eat, and where they live.

Pill bugs hold a special fascination for most young children. They can be picked up and observed. Although they, too, are scavengers and often live in the same place we find snails and slugs, the children will discover many differences.

Activities   1. In early morning, look for the silver trails left behind by snails and slugs. Look in damp areas and under overturned rocks for snails, slugs, and pill bugs. Observe where they live and how they move.

135

2. Collect a few snails, slugs, and pill bugs and examine them with a magnifying glass.

3. Create an indoor observatory (vivarium) for snails, slugs, and pill bugs by placing 2 inches of damp soil in a plastic shoe box. Collect several of the tiny creatures along with a few decayed leaves found nearby. Add a few small rocks. Sprinkle a small amount of water over the rocks and soil. Cover the shoe box with cheesecloth and tape in place. Observe for two or three days.

Explanation   Snails, slugs, and pill bugs are usually thought of as garden pests because they eat holes in the leaves on our plants, nibble on our strawberries, and ruin other fruits and vegetables. But they are helpful to the environment because their main source of food is dead and decaying plants and leaves. That is why we call them nature's clean-up squad.

A pill bug is sometimes called a sowbug. There are seven sections in the body of a pill bug. Because its body is broken up into so many sections, it can roll up easily. As a form of self-preservation, the pill bug rolls up when we touch or try to pick it up. Similarly, when we touch the snail, it draws in its feelers and moves back inside its shell. The silver trail left by snails and slugs is a special highway they travel on. Because they have no feet, they move about on their stomach and make this trail as they go from a liquid that flows from their body.

We learned how to make a temporary home or vivarium for our three scavengers. For a few days, they can live on the dead leaves and decayed plants we added, and they will have enough water from the moisture we provided. The children watched them crawl under the leaves and rocks because they all need a cool place to hide.

Follow-up
- Discuss how snails are born with thin, pink shells that grow as they grow. Explain that slugs do not grow shells.
- Examine the shell of a pill bug, watch it curl up when touched, and count the seven body segments and the seven sets of legs. These are just three of nature's scavengers who feed on decaying matter.
- Discuss the observations and then return the animals to their natural habitat.

Learning about the habits of insects

Goal To learn how to identify an insect and its specific uniqueness.

Teacher/Parent tips
- Locate places where insects can be found or collect several in separate jars. Make photocopies of pictures of insects that are found in your area.

Lab needs
- Insect books
- Crayons
- Insect picture sheets
- Plastic jars with holes punched in the lids
- Twigs, leaves, and small pebbles
- Black construction paper
- Plastic folders

Purpose The purpose of the following activities is to help children identify insects, and to give the little scientists a better understanding of the habits, food supply, and habitat of the ones they find in their area.

Children will discover that all insects have three body parts and six legs, and that spiders are not insects. They will be able to observe the differences and similarities in insects in both the indoor and outdoor labs.

By collecting and observing spider webs, the little scientists will discover what spiders eat and how they trap their food. They will also become aware of the different types of spiders and webs.

Activities
1. Look at pictures of insects in insect books. Locate and count the body parts and legs of various insects. Go for a walk and look for insects. Using a crayon, circle the insects on the insect picture sheets that you find. If available in your area, inspect milkweed plants in the spring for monarch butterfly eggs or larvae.
2. Collect insects in separate plastic jars to observe for a few hours and then release. Add a twig, leaf, and small pebbles to each jar.

139

3. Look for spider webs (early morning is the best time). Observe spiders at work. Collect spider webs by placing a piece of black construction paper behind the web and then pulling the paper quickly forward. The web will stick to the paper. Place in plastic folders. Collect spiders in jars for observation. Add two small twigs and watch for a web. Count the legs and body parts of the spiders that you caught.

4. Look for spider webs again. This time check closely to see where the webs are built and what types of insects are trapped in them.

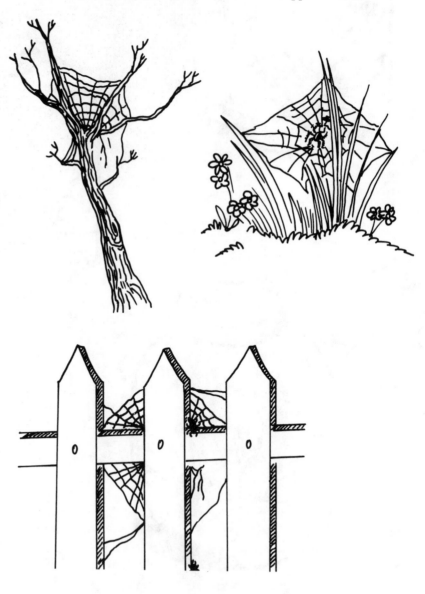

Explanation   There are thousands of types of insects. They all have three body parts and six legs; yet, each one is different, just as all of us are different. We have been observing many types of insects and found that some come in different colors, sizes, and shapes. Insects live all over the world, but not all types of insects live in the same place.

We provided air, food, and water for the insects that we kept in jars, but we should only keep them for a day or two because they need to be in their natural habitats to survive.

When we examined spiders, we saw that they were not insects. A spider has only two body parts and eight legs. Each type of spider looks different and builds its own special kind of web.

We did not harm the spiders by collecting their webs as spiders quickly build new ones. We discovered that spiders may build their webs between plants, across fences, just above the ground, and on top of grass. Spiders eat the insects they trap in their webs. They eat flies, ants, pill bugs, mosquitoes, and many other insects.

Follow-up   • Explain that all insects have three body parts and six legs.
• Discuss the different places where each insect was found.
• Explain that monarch butterflies lay eggs only on the milkweed plant because it is is the caterpillar's food supply.

- Discuss the uniqueness of each insect by referring to an encyclopedia or other reference source.
- Ask children to share what they have learned about insects.
- Compare the spider, which is not an insect, to the insects in the other jars. Note the differences and similarities. Explain how spiders spin webs by producing strands of silk from their tiny spinnerets and that they use the webs to trap their food. Study the web collection and discuss the differences.

Part 11
Stop! Look! Listen!—The Road to Discovery

This section presents suggestions for discovery-type walks that will encourage children to be more observant of the world around them. By focusing on one of the senses at a time, they will become aware of many concepts such as colors, shapes, and sizes.

Young discoverers will practice listening to sounds in nature and those made by people and machines, learn to identify these sounds, and try to mimic the sounds they hear.

Through specially planned walks, young botanist will be introduced to new smells and tactile experiences as they inspect flowers, trees, and plants.

The last section of this part covers the sun and shadows. Little scientists will have opportunities to track the sun on a chart that they make. Through these activities, they will learn that the sun gives off heat, that a shaded area is cooler than a sunny place, and that shadows can be seen on sunny days.

Learning about discovery walks

Goal To encourage children to take time to notice nature's colors, shapes, and sounds through heightened sensory awareness.

Teacher/Parent tips • Plan a series of walks to parks, around the block, to vacant fields, or around the play ground.

Lab needs • Shoe box with pieces of colored construction paper
• Plastic bag

Purpose The purpose of the following activities is not only to involved children in the healthy exercise of walking, but to make them aware of colors, shapes and sizes in nature.

  Children will discover that a "discovery walk" can be an outdoor laboratory when they take time to observe and learn about the world around them. By focusing on one concepts, such as color, they will become more aware of colors in nature, and see for themselves that not all trees have green leaves.

  Listening walks will help youngsters develop a sharper sense of auditory perception, as well as help them identify and classify sounds they hear.

Activities

1. Go for a "discovery walk" to see how many colors can be found in nature. When someone sees a color, "Stop!". Find the matching paper color in the box filled with various pieces of colored construction paper and place it in the plastic bag. Take turns calling out a color and have the others find the color in nature.
2. Go on a "discovery walk" and look for "shapes" in nature. Name the largest thing you can see. See who can find the smallest object. Take turns naming something you see when you look up; when you look down.

144

3. Go on a "silent" walk and listen for sounds in nature (e.g., a bird, frog, cricket, wind, rustling of leaves). When you hear a nature sound, hold up your hand for permission to share the sound. Try to identify the sound. Repeat, listening for first people-made sounds and then machine-made sounds.

4. Take turns mimicking the sounds that were heard from nature or machines. Have classmates try to identify the source of the sounds, such as a bird or a motorcycle.

Explanation We discovered that there are many colors in nature. We found trees with red, yellow, silver, and green leaves. This tells us that not all trees have green leaves. New leaves on trees and plants will not only be smaller but they may also be a different color than leaves that have been on the plant a long time.

Each type of tree, flower, and weed has its own shape and color. No two types of plants are exactly alike.

Follow-up
- Discuss what it would be like if there were no colors, just black and white. Count the colors in the plastic bag. Try to recall what things matched each paper color.
- Use the drawings of eyes and ears from an encyclopedia to show how each organ works.
- Share the shapes discovered in nature, the largest and smallest objects, and the different things we saw when we look up and down.

Learning about nature's sensory opportunities

Goal To make children more aware of textures and smells in nature.

New words Sun, shadow

Teacher/Parent tips
- Locate an area where there are several trees, flowers, and shrubs in bloom.

Lab needs
- Two paper bags for teacher

Purpose The purpose of the following activities is to encourage children to take time to smell, listen, feel and examine their world.

Children will discover during smelling experiences that not all things that are pleasing to look at have pleasant odors.

Memory and recall skills will be heightened through reinforcement games.

Activities
1. Go for a "silent" walk around the park or yard. Look closely at nature's gifts. Then take turns being the leader. Everyone close their eyes while the leader describes something they see in nature. Others try to guess what has been described.
2. Walk among the flowers and blossoms sniffing and smelling. Decide on which flower has the strongest smell, the faintest smell, the best smell, and the worst smell, and choose which is your favorite. Smell three flowers that are the same color and compare the smells.
3. Use your hands and fingers to feel dry soil, wet soil, and sand. Explore the texture of wet and dry grass, the top and back side of a large leaf, and the petals from three different flowers. Find one small item in nature and place it in the teacher's paper bag.

Hug a small tree. Hug a large tree. Feel the bark on both trees.

Explanation Nature is anything in the outside world that is not made by people. We discovered many of natures gifts on our discovery walk—sweet smelling flowers, tall trees, fuzzy seed pod, and colorful leaves.

Our nose told us that each flower had a special smell. Some smell nice and others don't. Because a flower is pretty to look at does not always mean it will smell good.

Each type of tree we examined had bark that felt different, and some tree trunks were so big we could not reach around them.

Follow-up
- Using the drawing of a nose found in an encyclopedia and explain how the nose allows us to smell.
- Discuss the different tactile experiences. Determine which tree the children could not reach around and how each trunk felt.
- Place one item at a time from the "collection" bag into the second, empty paper bag. Have children take turns first feeling and describing an object they found in the full bag, then guessing what it is.

Learning about sun and shadows

Goal To aid in developing an awareness of the sun and shadows.

Teacher/Parent tips
- Choose a sunny day for this group of activities.
- The sun is a star; the closest star to earth. It gives off light and heat.
- A shadow is made by anything that keeps light from going through it.

Lab needs
- Paper and crayons
- Sheets of newspaper
- Yarn

Scientific principle
- Anything that keeps light from going through it, will cast a shadow.

Purpose The purpose of the following activities is to help little scientists locate and record the position of the sun at three different times during the day. By experimenting, children will come to understand what a shadow is, what makes a shadow, and why shadows are longer at certain times of the day.

Children will learn to note the location of the sun at different times of the day, and will be able to give the direction of the sun at sunrise and sunset. Youngsters will be able to determine that the coolest place to be on a hot, sunny day is in a shaded area.

On sunny days, the little scientists will be able to see how the size of their shadow differs at different times of the day. By checking their shadow and marking the length, they will be able to see when the changes take place.

Activities
1. Have the children make an arc across a piece of paper with crayons. Check the position of the sun at three different times during the day and draw in the sun on the arc.
2. For approximately 20 seconds each, stand in the sun, in the shade of a building, under a tree where the sun is shining between the leaves,

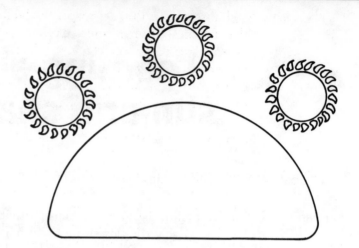

and in the sun with a newspaper "umbrella" over your head. Discuss the difference in temperature between the locations.

3. Find a place in the sun where your body will cast a shadow. Place a piece of yarn in front of your feet. Have a friend place another piece of yarn at the end of your shadow. Check your shadow at two other times during the day and mark the length of your shadow with yarn. Try to run away from your shadow. Find a place to hide from your shadow. Look for other objects that cast shadows.

Explanation   The sun rises in the east and sets in the west. Mark an "E" for east on your chart to show the morning sun and mark a "W" for west to show the late day sun.

   We found that standing in the sun without shade was the hottest place to stand, and that under a tree that had some sun shining through the leaves was cooler. The newspaper umbrella that we used kept the sun from shining on us so we felt cooler.

   Our shadow is longer in the morning and afternoon than at noon because, as our chart shows, the sun was lower on the arc. At noon it is almost overhead and our bodies do not block out the sun. This make our shadows shorter or almost disappear.

Follow-up   • Discuss the position of the sun on the charts that were made.
   • Share the differences that were discovered while standing in the sun, in the shade, under the tree, and under the newspaper umbrellas. Discuss why umbrellas may be useful on a hot, sunny day.
   • Compare the length of your shadow made at different times of the day. Share experiences and discoveries that are made when trying to run away or hide from your shadow. Discuss other objects that were found that cast shadows.

Sources

Science magazines

DOLPHIN LOG
The Cousteau Society
8440 Santa Monica Blvd.
Los Angeles, CA 90069

LADYBUG, THE MAGAZINE FOR YOUNG CHILDREN
P.O. Box 300
Peru, IL 61354

NATIONAL GEOGRAPHIC WORLD
National Geographic Society
17th and M Streets, NW
Washington, DC 20036

RANGER RICK
National Wildlife Federation
8925 Leesburg Pike
Vienna, VA 22184

SCIENCELAND, TO NURTURE SCIENTIFIC THINKING
Scienceland, Inc.
501 Fifth Ave., #2108
New York, NY 10017

3–2–1 CONTACT
Children's Television Workshop
One Lincoln Plaza
New York, NY 10023

Resources DOVER PUBLICATIONS, INC.
31 East 2nd Street
Mineola, NY 11501

  Large, full-color posters of plants and animals.

THE NATIONAL ASSOCIATION FOR
HUMANE AND ENVIRONMENTAL EDUCATION
Publishers of *KIND News*
67 Salem Road
East Haddam, CT 06423-0362

  Newsletter, audiovisuals, wildlife and environment information, leaflets, and posters.

NATIONAL WILDLIFE FEDERATION
8925 Leesburg Pike
Vienna, VA 22184

  Educational and environmental materials and books.

SEA WORLD
1720 South Shores Road
Mission Bay
San Diego, CA 92109

  Activity packets, curriculum guides, posters, booklets, and videos.

U.S. DEPARTMENT OF AGRICULTURE FOREST SERVICE
Woodsy Owl Campaign
P.O. Box 1963
Washington, DC 20013

  Record of noise pollution, poster, environmental reproducibles, song sheet, and leader's guide.